The Forever Dog

THE
FOREVER
Dog

SURPRISING NEW SCIENCE TO HELP YOUR CANINE COMPANION LIVE YOUNGER, HEALTHIER, AND LONGER

Rodney Habib and Dr. Karen Shaw Becker

with Kristin Loberg

HARPER

NEW YORK · LONDON · TORONTO · SYDNEY

HARPER

A hardcover edition of this book was published in the United States by Harper, an imprint of HarperCollins Publihsers.

FOREVER DOG. Copyright © 2021 by Planet Paws Media, Inc. All rights reserved. Printed in the United States of America. No part of this book may be used or reproduced in any manner whatsoever without written permission except in the case of brief quotations embodied in critical articles and reviews. For information in the U.S., address HarperCollins Publishers, 195 Broadway, New York, NY 10007, U.S.A.

HarperCollins books may be purchased for educational, business, or sales promotional use. For information, please email the Special Markets Department in the U.S. at SPsales@harpercollins.com.

FIRST U.S. HARPER PAPERBACK EDITION PUBLISHED 2025.

Illustration on page 27 courtesy of the Israel Museum, CC0, via Wikimedia Commons. This image was taken at the Israel Museum in Jerusalem, Israel, by Gary Lee Todd, PhD, professor of history, Sias International University, Xinzheng, Henan, China. He has dedicated the work to the public domain and it is available under the Creative Commons CC0 1.0 Universal Public Domain Dedication.

Designed by Leah Carlson-Stanisic

Library of Congress Cataloging-in-Publication Data has been applied for.

ISBN 978-0-06-300261-6 (pbk.)

24 25 26 27 28 LBC 5 4 3 2 1

To Sammie, Reggie, and Gemini,
our first teachers

Contents

Part III Pooch Parenting to Build a Forever Dog

Authors' Note

This book is heavily referenced with citations to primary source material, secondary information, and links to further resources. But you won't find any hint of them between the covers of the actual book. Why not? Well, there are just too many to include. It's an avalanche. Hence, we decided to house them online at www .ForeverDog.com to keep the cost and length of this book down for you, our dear reader. It's a win-win for everyone because we get to save a few trees and render the book more affordable and accessible, and the bonus is we can keep our references as updated as possible in dynamic fashion—adding more as new science emerges. We've left no stone unturned in this department, including the sound science and historical data to defend even our boldest assertions and statements that buck established dogma. Everything in this book, including claims that may seem outrageous, can be backed up by indisputable evidence. The reference list is your magic key to debunking the disinformation in the pet care world, learning the real true science, and equipping yourself with the tools to support a Forever Dog.

INTRODUCTION:
Humanity's Best Friend

We are all just walking each other home . . .

—Ram Dass

. . . and we hope it will be a really long walk.

—Dr. Becker and Rodney

Dr. Curtis Welch was worried. As winter began its slow creep into the small Alaskan town of Nome in late 1924, he noticed a disturbing trend: increasing cases of tonsillitis and inflammatory diseases of the throat. The flu pandemic of 1918–1919 that killed a little more than a thousand in his state was still fresh in his memory, but this was different. Some of the cases had the appearance of diphtheria. In his eighteen years practicing there as a physician he had never seen the contagious infection—brought about by a certain strain of toxin-producing bacteria—that can lead to death, especially in children. Diphtheria was commonly known as the "strangling angel of children"; it causes the throat to become blocked with a thick, leathery coating that makes breathing very difficult. Without treatment, death by suffocation is likely.

By January the following year, it was clear he was dealing with an outbreak of the dreadful disease and had no treatment on hand. Children were beginning to die. At his request, all schools, churches, movies, and lodges were closed, and all public gatherings were forbidden. No longer could people travel on the trails, with the exception of those delivering mail or conducting urgent and necessary business. Homes were quarantined if any family members were suspected of having the illness. Although these measures would help, what Dr. Welch really needed to save his entire region of about ten thousand people was the antitoxin serum. The cure, however, was more than one thousand miles away in Anchorage; it may as well have been millions of miles away, as there was no way to traverse the ice-filled harbors or fly an open-cockpit plane through the subzero temperatures.

Enter teams of sled dogs and their drivers to participate in the Great Race of Mercy, a historic round-the-clock relay that covered 674 miles of rugged wilderness, frozen waterways, and treeless tundra over five and a half days to deliver the miracle serum to Nome. Two Siberian huskies named Balto and Togo stood out as the superstar dogs of the journey. They often relied on scent rather than sight to push through miles of blinding, whiteout conditions—a perilous trip that is now part of the iconic Iditarod Trail. This story is just one of many that vividly illustrates how incredible dogs are and how humans and dogs have been helping one another since we fell in love thousands of years ago.

It has been nearly a century since the serum run saved Nome, and, as the irony of the world's stage would have it, we write this while in the midst of another outbreak circling the globe. Society is searching for our modern version of such rescue dogs to save us from an invisible foe that has proven lethal in many people. A sled dog is not likely going to deliver the antiserum today (though we wouldn't put it past some more sled dogs coming to the rescue in remote areas again for delivering anti-COVID therapies and vaccines), but our dogs have no doubt become central players in antidotes of other kinds that are helping us plow through the coronavirus pandemic. More than half of US households own pets, with dogs taking the lead over cats. By some estimates, 12 percent of adults with children under the age of eighteen adopted pets because of the pandemic, compared with 8 percent of all adults. Pet owner-ship is trending upward, and we think it's here to stay.

For many dog owners,* our pets provide brief moments of para-dise on long, refreshing walks and are regular sources of hugs and kisses in our homes. They provide unwavering comfort, cuddles, and unconditional love—a distraction from the bad news and a bea-con of hope for tomorrow. In some small communities, "winery" and "brewery" dogs are now delivering libations, and scientists are training a few breeds to sniff out sick people so they can be used at airport checkpoints.

The experience of this pandemic has highlighted the importance of dogs in our lives and the role they have in helping humanity move along and, as it were, *survive.* As much as they depend on us for life's necessities, we rely on them for an untold number of things. They ultimately help us be better people physically, mentally, emo-tionally, and, one can easily argue, professionally (numerous com-panies now list an office dog as an employee). It's really true that owning a dog has been proven to extend the longevity of humans. A growing body of evidence links dogs to good health, and not just for the obvious reasons of lowering our general stress levels and feelings of loneliness. Studies show dogs can lower our blood pressure, keep us active, reduce the risk of heart attack and stroke, boost our self-esteem, encourage social engagement, force us to be outdoors in nature, and trigger the release of potent chemicals in our bodies that make us feel safe, connected, and content. One study even revealed that dog ownership is linked to a 24 percent reduced risk of dying *from anything* (what the scientific literature likes to call "all-cause mortality"). And for people with underlying

---

* We realize that people use different terms to refer to their relationship with their pet—"pet parent," "owner," "guardian," and so on. Some may take issue with the term "pet" or "dog owner," but there's no general consensus on this matter, so you can choose whichever terminology you prefer. We'll mix up the language in the book.

cardiovascular conditions, which amounts to millions of people in the United States alone, that reduced risk is even greater. In 2014, Scottish scientists calculated that owning a dog, particularly later in your life, can turn back the aging clock and make you act—and feel—ten years younger. We've also learned that dogs can help children develop stronger immune systems and soothe the stressors of adolescence, a period of time often riddled with self-doubt, peer judgment, adult expectations, and emotional turmoil.

Dogs serve us well in many ways, from helping us to keep better schedules (after all, they must be fed and walked on cue) to protecting our families and sensing danger. They can detect the imminent doom of an earthquake minutes away and smell environmental changes in the air that signal a major storm or tsunami coming. Their keen senses make them excellent helpers in tracking down criminals, finding illegal drugs and explosives, and locating people who are trapped or, worse, dead. Their olfactory prowess can nose out cancer, dangerously low blood sugar in a diabetic, pregnancy, and now COVID-19. And they can be surprising wellsprings of thought and inspiration. It has been suggested by scholars that dogs in fact introduced Darwin to the systematic study of nature and came to help shape his scientific approach in his formative years. (Polly, a smart terrier, was often found snuggled in a basket by the fireplace near Darwin's desk in his study where he wrote his seminal masterpiece *On the Origin of Species.* They'd converse in front of the window, with Polly barking while Darwin jokingly referred to "the naughty people" outside. Polly became a model for illustrations in Darwin's last book, 1872's *The Expression of the Emotions in Man and Animals.*)

But not all the news is pooch-perfect rosy. In our lifetimes alone, we've often seen the lifespans of canine decrease, especially among pedigree dogs. We realize this is a bold and controversial statement to make, but bear with us. Although many dogs are indeed living longer, like people, many dogs are dying prematurely of more chronic disease than ever before. Cancer is the leading cause of death in older dogs, with obesity, organ degeneration, autoimmune disease, and diabetes not too far behind. (Younger dogs are more likely to die of trauma, congenital disease, and infectious diseases.) We have met countless pet parents who are desperate for ways to keep the dogs in their lives for as long as possible (maybe not "forever," but at least as long as they can for a healthy life span, or health span; "life span" and "health span" are two important terms that are not one and the same; we'll distinguish them shortly).

We should be clear from the get-go: We're not aiming to teach you how to have a dog that literally lives forever. And we're not going to solve every dog's health problems in this book either—there are too many variables and potential permutations of different health conditions across all types of dogs to be able to spot-treat every possibility (but you can find a gateway to personal pet problem solving on our website at www.foreverdog.com). The purpose of this book is to offer a science-backed framework for optimal dog parenting and caring that you can tailor to your unique circumstances. And the purpose of calling this book *The Forever Dog* is equal parts metaphorical and aspirational. We aspire to have dogs that live vibrant lives to the very end—whenever that is. In death, they stay with us, too. After all, our dogs are forever in our hearts and minds, even when they are gone from this planet. Your dog has found a forever home with you, and you want to make the most of it.

Forever Dog (fə-ˈre-vər dȯg): A domesticated carnivorous mammal, descended from the gray wolf lineage, that lives a long and robust life free from degenerative disease, in part due to their humans' making intentional choices and wise decisions that confer health and longevity.

Interestingly, we didn't really consider our pets family members until after World War II. In 2020, historical geographers who analyze the changing relationship between humans and animals finally figured this out by looking at gravestones in Hyde Park, England, which date back to 1881. It holds a secret pet cemetery. After collecting data on 1,169 grave markers from 1881 to 1991, they discovered that before 1910, only three gravestones—less than 1 percent of those surveyed—referred to a pet as a family member. But after the war, almost 20 percent of grave markers described pets as family, and 11 percent used surnames. The zooarchaeologists, as these researchers are called, also noticed more cat graves as time went on. And in 2016, for the first time, New York made it legal for pets to be buried with their owners in human cemeteries. If our pets deserve a spot in heaven with us, then they deserve as good of a spot—and life—on Earth with us, too.

Our mission is to empower the tens of millions of dog owners, and anyone hoping to become one, to change the way they care for their pets in order to improve the wellness of their animals, maintain their vibrant health, and ultimately extend the life span of dogs worldwide. They deserve to be free from chronic disease, degeneration, and disability. These are not inevitable outcomes of age alone (even in us!). But reaching this goal requires a change in thinking as we take you on a vivid, science-backed tour of all the key factors that can help lengthen dogs' life spans. Although we do get into the details of the science, we promise to make it easy. The research we've included in this book is meant to educate and inspire you, and to provide the data and background for you to feel comfortable making the important lifestyle changes needed to maximize your dog's health and longevity. Depending on how familiar or foreign some of these wellness concepts are to you, our recommendations may seem overwhelming, so we've provided lots of options for small steps that you can incorporate into your dog's routine in whatever manageable increments your brain's bandwidth, your schedule, and your budget allow. We field questions daily from a wide array of people who seek tips and solutions to maximize their pets' quality of life, and we know that our audience is as broad and diverse as it is deep and devoted. We come from different backgrounds and experiences, but we all have the same goals in mind when it comes to our dogs.

We aim to strike a balance between speaking to those who need some hand-holding and those who crave the scientific intricacies. If you don't understand something, move on and don't worry about it; the commonsense strategies will make sense by the end of the book. We trust you'll gain a lot of knowledge, even if you just skim the heavy science, and we'll be doling out practical tips along the way. We would be remiss to avoid describing the biology behind the fascinating facts of our dog's existence (and our own, too). We also would be irresponsible to dodge difficult and sensitive conversations. For example, we're all reminded, whether we like it or not, that weight is a major issue in health circles today. Talking about those extra pounds is a taboo topic that many doctors—veterinarians included—don't like to have in their offices. It's unpleasant and awkward. It's the third rail that's borderline shame-raising. But the conversation is necessary. We're not casting blame; again, we're offering solutions. When excess weight adversely affects health, it's like running while clumsily carrying a sharp object. None of us would let our dog run with a blade in his mouth, right? If there's one lesson you're going to learn over and over again, it's this: **Eat less, eat fresher, and move more and more often**. That's a truism for both you and your dog. And it's the biggest takeaway you're going to get from this book. Even though we just gave away the whole point of the book in ten words, you can't close this book yet because you need to know *how* and *why* you and your pooch are going to eat less and fresher and move more. When you know *how* and *why*, the *doing* takes care of itself.

We live in exciting times, thanks to the speed of technology and acquired wisdom about the mammalian body over the past century. Our understanding of activities inside cells has been exponentially increasing, and we are excited to present this new knowledge within the context of one fine goal: helping our darling dogs thrive as best they can alongside us.

Many of our lessons debunk commonly held myths and practices, especially when it comes to diet and nutrition. Like many humans, lots of dogs are overfed and undernourished. You know that eating highly processed food at every meal is probably not a good idea. That's pretty obvious. What's not commonly known is that most commercial pet food is just that—highly processed fare. Please don't feel too shocked or duped. You're not the only one in the dark. But the news isn't all bad because—just like the processed foods you enjoy (ideally in moderation) in your own life—you don't have to stop feeding your dog commercial pet food entirely. You can make it work in moderation with the guidance we provide, and you get to choose at which level you want to replace commercial pet food with fresher options.

Fresher Is Better: "Fresher" dog foods, no matter the form (homemade, commercial raw or gently cooked, freeze-dried, or dehydrated) promote longevity far better than modified, ultra-processed kibble and canned pet food. We'll show you how to tweak your dog's diet to incorporate more of them into his daily grub.

As it's said, it starts with food but doesn't end there. Many dogs are also deprived of adequate exercise while at the same time saddled with the impact of environmental toxins and the effects of our own toxic, unrelenting stress. We'll also discuss the ways you can begin to understand your dog's genetic past and present and use this information to leverage proactive caregiving to mitigate the effects, for instance, of less-than-optimal genetics.

Breeding practices over the past century have radically altered many of our dogs—some for the better but unfortunately a lot for the worse. Sure, domestication brought floppy ears and more docile genes, but rampant and unstructured breeding also brought out recessive genes and created genetic deletions and narrow gene pools. Such circumstances have contributed to "breed flaws" that generate genetically weak animals. One in three Pugs cannot walk properly due to "vanity breeding," which leads to, among other things, a higher risk for lameness and spinal cord disorders. Seven out of ten Dobermans carry one or both genes for dilated cardiomyopathy because of "popular sire syndrome," which occurred decades ago. (Dilated cardiomyopathy, or DCM, is a condition in which the heart loses its ability to pump blood because its main pumping chamber is enlarged and weakened; it's a common type of heart disease in humans, too.) The good news is there's a lot we can do to change this picture. Dogs are our canaries in the coal mine (or "canines in the coal mine"); their health struggles in the past fifty years parallel ours. They age similarly to us but much more rapidly, which is why research scientists are increasingly respecting dogs as models of human aging. Unlike us, however, our pets can't make health decisions for themselves. It's up to pet parents (or owners or guardians or whatever you prefer to call yourself in your dog's life) to make smart choices for lasting vitality and health. We'll show you how and make this as practical and doable as possible.

In Part I, we'll discuss the modern age's dramatic decline in the number of healthy dogs, couching our conversation in a breathtaking view of our co-evolution. Dogs may have exploited a niche they discovered in early human society, when they persuaded us to bring them inside, shelter them from the cold, and feed them.

Put another way, the dogs warmed up to us rather than the opposite. They entrusted themselves to our care, and we accepted and embraced the challenge. In these chapters we now sound the alarm, describing the full spectrum of challenges dogs face today in their health and wellness, largely as a result of the trust they placed in us, their human caretakers, and we hint at solutions.

In Part II, we'll delve into the gems from science and what we know about de-aging through diet and lifestyle. You'll learn how food speaks to genes, why a dog's gut germs (microbiome) are just as important as our own inner microbial world for health, and why it's important to cater to your dog's choices (at least occasionally) and honor her individual preferences. Finally, in Part III, we unveil our Forever DOGS Formula and show you how to implement the strategies in real life to build a Forever Dog. We'll be equipping you with all the tools you need to tailor our recommendations to your companion's life and essentially stack the deck in favor of your dog's robust and vibrant health span. We predict that you will be transformed as well. You'll start to think about what you're eating, how much you're moving, and whether or not you're living in an environment conducive to health.

THE FOREVER DOGS FORMULA

- ➤ **D**iet and nutrition
- ➤ **O**ptimal movement
- ➤ **G**enetic predispositions
- ➤ **S**tress and environment

To make things really simple and practical, we'll end each chapter with Longevity Junkie Takeaways and suggestions for things you can consider doing today. Along the way, we'll put key phrases to remember in **bold** and highlight certain facts in boxes. You won't have to wait to get to the in-depth How-Tos we outline in Part III; we'll give you actionable information from the very beginning so you can start to make small but meaningful changes right away. Again, we reflect what you ideally want from this book: answers to the *why* from the science and the *how* from the everyday reality of putting that science into practice.

Our animal-loving audience is diverse. If you're new to proactive living, we're hoping this is the beginning of a long and healthy friendship, centered around all we can do to maximize our dogs' well-being as they age. Our core community is made up of hundreds of thousands of "2.0 pet parents"—empowered, knowledgeable guardians using deliberate, commonsense approaches to creating and maintaining vibrant health for their animals. These committed pet parents have been using innovative health strategies for the last ten years (and many of them for much longer) and have begged us to put the longevity and wellness wisdom into a reference book so that their veterinarians, friends, and family can read the pertinent science in one place. We also meet a lot of brand-new dog owners revamping their lifestyles (including how they approach caring for family members). Our goal is to include enough background information so the recommendations make sense to the newcomers. We also wanted to provide the leading research for our Longevity Junkies—the biohackers always looking to tweak day-to-day choices to optimize their dogs' health. We don't want this information to be overwhelming for the people who are brand new to the concept of proactive health; our goal is to inspire, so take one tip at a time and integrate it into your dog's life in a way that works for you.

> **Longevity Junkie:** Someone who seeks everyday secrets to extending health span—the time spent in life free of disease, disorder, and disability—with basic lifestyle strategies.

The two of us joined forces a few years ago, and you'll come to learn about our individual and collaborative journey in the pages ahead. As two dog lovers devoted to helping pet owners navigate the confusing world of animal health, we had no idea just how aligned our goals were when we kept running into each other at conferences and lectures. No sooner did we connect and forge a professional partnership than we realized there was an opportunity for us to achieve a shared dream together: to retrain the collective mindset regarding dogs and their health. We knew the task would be challenging; in fact, over the course of the last several years, we've traveled to many parts of the world gathering the most up-to-date information about dog health, disease, and longevity. We interviewed top geneticists, microbiologists, oncologists, infectious disease doctors, immunologists, dieticians and nutritionists, dog historians, and clinicians, collecting cutting-edge data to help our mission. We also interviewed people who have owned the oldest dogs in the world to find out what they did—or didn't do—to enable their dogs to live into their twenties and, in some cases, even into their thirties (the equivalent of a super-aging human living to one hundred ten or older, what's called a "supercentenarian"). What we discovered has the potential to revolutionize the pet world forever. The information we've amassed, much of which we hope will astound and motivate you, will increase longevity in beloved pets everywhere. And perhaps even in you, too. As we like to say: "Health travels up the leash."

Veterinary medicine is as much as twenty years behind human medicine. The latest senolytic (antiaging) research will eventually trickle down to reach our pets, but we don't want to wait. There are also critical aspects of canine health that aren't part of mainstream conversation, but that's changing, thanks to One Health—an approach that recognizes the health of humans is intimately connected to the health of animals and our shared environment. The One Health Initiative isn't new but has become more important in the last few years as physicians, osteopaths, veterinarians, dentists, nurses, and scientists recognize that we can learn more from one another through co-equal, all-inclusive collaborations. "Collaborative efforts of multiple disciplines working locally, nationally, and globally to attain optimal health for people, animals, and the environment" defines the One Health Initiative. So while research that unites human and veterinary medicine may not be mainstream quite yet, it's coming. We cover a lot of human health science in this book because it's the basis of much of the companion animal research in the works, and vice versa.

The One Health concepts and correlations we discuss in this book are not being widely broadcast in pet webinars or magazines; they are not discussed by most veterinarians and pet owners; and they are not popping up on social media . . . *yet*. We want to start a much-needed conversation—and to spark a movement. The basics of what makes humans healthy (or unhealthy) also apply to dogs; we'd like to get that conversation started right now.

> **Editorial Note:** We write this book from a collective "we" as authors, but occasionally we will break into our individual voices (Rodney or Dr. Becker), and when we do that, it will be clear who is speaking. We also freely interchange "he" and "she" when referring to your dog.

The physical and emotional well-being of our dogs is shaped by the choices we make for them. And the well-being of our dogs in turn impacts us. That leash is a two-way street. For centuries, humans and canines have enjoyed a symbiotic bond, with each affecting and enriching the life of the other. As the world of medical research becomes more global, the choices for canine health are as vast as those for human health. We all must choose wisely, to raise a Forever Dog.

PART I

The Modern, Unwell Dog

A Short Story

Sick as a Dog

Why We and Our Companions Are Losing Our Longevity

> The reasons for some animals being long-lived and
> others short-lived, and, in a word, causes of the
> length and brevity of life call for investigation.
>
> —Aristotle, "On Longevity and Shortness of Life," 350 BCE

Reggie was a dog destined to be a Forever Dog, at least in our minds. At ten years old, the Golden Retriever was thriving. He had never suffered an ear infection, never needed his teeth cleaned, never had allergies or hot spots, or experienced *any* of the midlife or senior symptoms that plague so many dogs. His body worked perfectly, visiting the vet only every six months for genuine "wellness exams." Reggie's biannual blood work was also perfect, including his cardiac markers to check for heart problems. Reg didn't have a single health issue in his entire life, and what a good one it was with Rodney as his dad. On December 31, 2018, Reggie refused to eat breakfast, a clear sign something was wrong. Within two hours he had crashed: cardiac hemangiosarcoma, a cancer of blood vessels around the heart. The shift from healthy to deathly was so sudden, so unexpected and traumatic. He was gone within a month.

The pain of Reggie's shocking departure was compounded by

the fact that Sammie, Rodney's White Shepherd, was the dog everyone knew would die of genetic disease. Someday. Sam was diagnosed with degenerative myelopathy four years prior, the dreaded heritable disease that paralyzes its victims, starting in the rear limbs. Sam defied all odds, successfully battling her diagnosis and maintaining the use of her body, thanks to intensive daily physical therapy and innovative neuroprotective protocols that were instituted immediately after her diagnosis. But that changed the day Reggie died. Sam and Reggie were best friends, and the day Reggie passed it was clear Sammie gave up, too. Sam's condition declined rapidly after Reggie's transition, resulting in double heartbreak for Rodney.

Losing Reggie and Sam stopped Rodney's life, as death has the power to do. Irreplaceable loss doesn't just knock you off course and bring you to your knees, it makes you not want to continue. The grief is especially difficult if the loss was unexpected or premature. But loss is loss. Grief counselors and therapists acknowledge the loss of a beloved animal companion can be equal to the loss of a human. There's a definite "before" and "after," with no going back. In these moments most of us come to one of two conclusions: *I will never do this again; it's just too painful.* Or option two: *If I do this again, I will be wiser, I will know more, and I won't let that happen again, at least not in the same way.* If you fall into the second camp, this book is for you.

Writing this book has been a form of therapy for Rodney and a personal evolution for both of us, especially when it comes to how we look at genetics. Losing dogs to genetics isn't something most people think very much about, especially when staring at a fuzzy eight-week-old yummy puppy. When you fill out the background paperwork for your new vet visit, you don't see the same questions asked on a new human patient form at the doctor's office (i.e., What did your paternal grandparents die of? What did your maternal grandparents die of? Is there a history of cancer in your family? Have any of your siblings been diagnosed with certain diseases?). Getting answers to these questions would be completely eye-opening in the veterinary realm—and impossible. It would shed light on the fact there have been profound and detrimental changes in our dogs' genomes, and in a relatively short period of time (considering).

The cancer that struck Reggie is more common among Goldens than other breeds largely due to how they've been bred. Most modern Golden Retrievers are saddled with genes that predispose them to certain cancers; similarly, Chocolate Labs live shorter lives than other Labs—10 percent shorter—due to selective breeding for the color coat that simultaneously introduces harmful genes. How genetics, lack of genetic diversity, genetic deletions, and genetic mutations play into your dog's overall health and disease are stand-alone books with deep science. We just want to give you the rules of the road for avoiding, whenever possible, the genetics-related heartbreak that we have experienced. For those of you who spend money on puppies (that means not adopting or rescuing), boy, oh boy, do we have a long list of questions for prospective breeders requiring *incredibly satisfactory* answers, before you spend a dime. If you're going to spend money on a dog from a breeder, spend money on stellar genetics.

For those of us who have no idea what our dog's genomes hold (and may never know, for whatever reason), or who have a genetically damaged dog from a puppy mill, don't dismay. We've interviewed some of the top canine geneticists in the world, all of whom say the same thing: There's hope, *epigenetically*, to help these dogs maximize their health span, despite terrible genetics. While we can't change our dogs' DNA, **there is a massive amount of research demonstrating our ability to positively influence and control gene expression**, and that's what this book is about. We'll be going into the magic of epigenetics soon.

Our job as guardians is to maximize our dogs' health in order to extend their life spans by removing all of the roadblocks we can. Our goal becomes helping them live every day to the fullest with a high-quality existence.

But why are they getting deprived of a life free of disease and dysfunction in the twenty-first century, when we have more veterinary knowledge than we ever have had before? Granted, in general, humans will always have longer life spans than dogs. But we should not accept the heartache of witnessing a dog's premature death, sometimes over and over again throughout our lifetime with different dogs. Can this be changed? For genetically damaged dogs, if we can't lengthen the time our dogs are with us, can we dramatically improve the quality of their existence while they're here? Can we defy some of the odds against them? The answer is most definitely yes. Even dogs that have won the genetic lottery and don't have underlying genes that can spell disease or dysfunction are vulnerable to premature death today. This, too, can be remedied with an understanding of *why*. First, we need to look at what's going on among dogs' favorite counterpart: humans.

The Extinction of the Healthy Dog

The ancient Greek philosopher and scientist Aristotle was ahead of his era. Although most of us think of him as a purveyor of lofty, esoteric wisdom about ethics, logic, education, and politics, he was also a polymath on natural science and physics and a pioneer in the study of zoology, both observational and theoretical. He even wrote about dogs and their various personalities, noting with admiration the longevity of Odysseus's faithful dog Argos in Homer's classic epic poem. When Odysseus returned to the kingdom of Ithaca after ten years fighting at Troy and another ten years struggling to get home, he arrived disguised as a beggar to test the loyalty of his friends and family. Only his old dog Argos recognized his master, greeted him rapaciously with a wagging tail, and soon thereafter died happy. Argos had lived well into his twenties.

The mysteries of aging have been debated for more than two thousand years. Aristotle was not entirely right to think aging had something to do with moisture (according to his reasoning, elephants outlasted mice because they contain more liquid and thus take longer to dry up), but he was right about a lot of other things and set the stage for modern schools of thought.

If we asked you what you should be doing to maintain your youth, to live a healthy life span during which you're active, free of disease, and able to avoid the unwanted side effects of aging, what would you say? Perhaps it's some or all of the following:

➤ Prioritize good nutrition and regular movement to sustain ideal weight, metabolic health, and physical fitness
➤ Achieve restorative sleep nightly
➤ Manage stress and anxiety (with the help of a dog)
➤ Avoid accidents, exposures to cancer-causing agents and other toxins, and deadly infections
➤ Stay socially active, engaged, and cognitively stimulated (i.e., keep learning)
➤ Choose parents with longevity genes

Obviously, that last idea is beyond your control, but if you aren't one of the individuals who was born with perfect genes (there's no such thing), you might be relieved to know your genes account for much less than you'd think when it comes to your life span. Science finally figured this out, thanks to analyses of large ancestry databases only recently available. In the past, it was believed that 20 to 30 percent of genes contribute to a person's life span, but new evidence shows it's well under 7 percent. That means that the majority of your longevity is in your hands and based on your lifestyle choices—what you eat and drink, how often you break a sweat, how well you sleep, what kind of stress wears on you (and how you cope with it), and even other factors like the quality and strength of your relationships and your social networks, whom you marry, and your access to health care and education.

In a 2018 study of the life span of spouses, which entailed family trees of more than 400 million people who were born from the nineteenth century to the mid-twentieth century, scientists at the Genetics Society of America could make these newer calculations. They found that married couples shared similar life spans—more so than those of siblings. Such an outcome suggests a strong influence on nongenetic forces because spouses do not typically carry the same genetic variants. Other factors that they likely do have in common, however, include eating and exercise habits, living far from disease outbreaks, access to clean water, literacy, and non-smoking status. This makes sense: People tend to select partners who share a similar lifestyle. You rarely see a smoking couch potato hooking up with a competitive, smoke-free fitness buff. We prefer to spend our lives (and procreate) with people who love what we love, whether in ideology, values, hobbies, or habits. This phenomenon actually has a name: assortative mating. We tend to select mates who are similar to ourselves.

We all want to live as long as possible in a healthy state. Most antiaging researchers are not seeking immortality. Our guess is neither are you. What we all want and aspire to, however, is an *extension of the health span*; we want to add a decade or two to our vibrant, joyful days and shorten the period of time we spend in "old age." In our dream world, we die peacefully in our sleep "of natural causes" after our last wonderful dance. There's no pain. No chronic illness to manage for years or decades. No reliance on powerful drugs to get through the day. We want the same for our pets, too. The good news is that if the information is put into action, **science likely already knows enough about the biology of aging to increase our dogs' health span by three to four years**. That's a long time in dog years. We can't make any guarantees, but we can say with confidence that if you put certain proven strategies into practice, you're helping your dog increase his or her odds of earning those bonus years.

"Squaring off the curve" ("squaring the mortality curve") is a way to look at extending longevity. It means that your morbidity

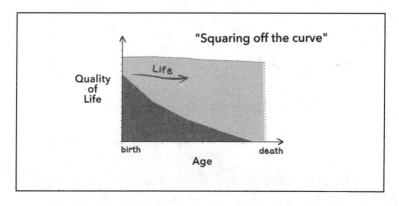

risk (chance of death) remains low as you grow old; rather than your becoming increasingly frail as you age, your good health lasts right up until a short time before your death. "Happy healthy happy healthy happy healthy dead" is how we'd like to live (and die). This is in stark contrast to what we're conditioned to believe will occur (the downward-sloping dotted line): By midlife or certainly retirement, we will have myriad physical symptoms that impact our mobility and/or brain function; we will be prescribed a growing list of medications to manage our degenerating bodies; and then we will develop cancer or Alzheimer's disease, have a heart attack, stroke, or organ failure, wallow for a while, and then die. Ick. Science says we have a significant impact on which of these two scenarios plays out, based on our lifestyle choices. But what about our dogs? They can't make the best choices for themselves because we're in control. And there's currently no blueprint for creating a long and well-lived life for your dog, which is why we are so passionate about the work we do.

By combining the pearls of wisdom we've gleaned from studying the oldest dogs in the world with the most up-to-date longevity research and emerging translational science, we hope to empower you with the knowledge you need to make wise decisions for your canine friend. By making consistent, well-informed lifestyle choices for your dog, you'll move away from high-risk variables and early

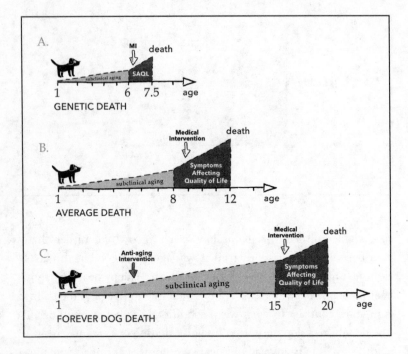

degeneration because you'll be taking steps designed to avoid them. Statistically this results in a longer health span.

Obviously, some of the factors in human longevity do not apply—dogs don't earn degrees, smoke cigarettes, or get married. What's more, for some dogs, as we'll later see in detail, genes can factor a little more heavily into their longevity equation. But let's cast the genetic component aside for a moment, because the power of environmental forces eclipses purely genetic forces. After all, as we'll cover in a later chapter, genes behave *within the context of their environment,* and there's a thought experiment here worth pursuing. Dogs do indeed share a lot with us. They live in our houses, breathe our air (and our secondhand smoke), drink our water, follow our direction, sense our emotions, eat our food, and sometimes even sleep in our beds. It's hard to imagine an animal that shares the human environment more than dogs. It might help to picture for a moment what it must be like to be someone's beloved pet and at their mercy (but to your delight):

You're fed and taken out on walks at regular intervals; you're bathed, combed, kissed, and cuddled. You've got a favorite spot for catching an afternoon nap; you've got your favorite toys and places to sniff and poop. You have friends in the park and love to play with fellow canines as well as your owner. You especially love getting dirty outside, exploring new places, sniffing another canine's butt, and having novel interactions with others.

Such an image might remind you of your childhood, when you were fully reliant on an adult to take care of you, clean up after you in more ways than one, and overall ensure your safety. Though you could have protested in some way, you didn't really have a say in what you ate, when you were bathed, or how many times you were taken to the park or playground. But you went along with it because that's all you've ever known. You have an instinctual, arguably inborn trust in your parent or guardian. And you grow up with habits that are shaped by that upbringing. Today, as an adult, you probably owe a great deal of your health (or lack thereof) to your everyday habits, whether they support a good long life or drive you in a direction of chronic illness.

For most of us, we mature to become independent and can choose to modify our habits to fit our needs and preferences as we age. But for our dogs, they rely on us for life. In the whole scope of life, we give our dogs very few choices. When illness strikes, we have to ask: *What went wrong?*

It's commonly known that we humans suffer increasingly from the so-called diseases of civilization, such as diabetes, heart disease, and dementia, that are largely brought on over time by lifestyle choices (poor diets, lack of exercise, and so on). A slow-moving tsunami builds over years or decades before reaching our biological shorelines. Although we may be living longer than we did a century ago in large part owing to improvements in nutrition, sanitation, and drug development, are we living *better* longer?

In 1900, according to the World Health Organization, the global average life expectancy was just thirty-one years; even in the wealthiest countries it was under fifty (in the United States it was about forty-seven years old). We should temper the significance of those numbers, however, because "average" life expectancy in the early twentieth century was dragged down by the toll of infectious diseases, especially among children, that resulted in premature deaths. Once antibiotics became widely available and we learned how to treat a lot of illnesses, average life expectancy increased significantly. By the twenty-first century, the main causes of death and disability had shifted from infectious diseases and infant mortality to noninfectious ailments—or chronic diseases—in adulthood.

By 2019, before the pandemic hit to skew the numbers, the average life span was nearing seventy-nine in the United States, and it was as high as eighty-four and a half in Japan. But get this: Fewer than 50 percent of people who live in the States today makes it past eighty, and two-thirds of those will die of cancer or heart disease, with many of the "lucky half" who make it past eighty succumbing to sarcopenia (loss of muscle tissue), dementia, or Parkinson's disease. What's more, we've lost gains in our life expectancy more recently beyond the effects of the COVID-19 pandemic, with numbers showing a slowdown (and, by some measures, a total halt) in our ability to lengthen quality of life. We have made huge strides in the last century to raise the bar on life expectancy, but we now face a higher, largely self-imposed bar in our efforts to extend healthy life. Age inevitably causes wear and tear on our bodies, but we increasingly succumb to conditions that are highly avoidable and that eventually anchor themselves into chronic, intractable illnesses.

It doesn't have to be this way. Cancer, heart disease, metabolic dysfunction (think insulin resistance and diabetes), and neurodegenerative diseases like Parkinson's and Alzheimer's are still rare in many parts of the world, including small regions of even modernized countries. In certain regions known as Blue Zones, as many as three times the number of people reach one hundred and beyond, retaining their memories and good health much longer than we do.* In 2019, one of our most prestigious medical journals, *The Lancet*, published an alarming study stating that we now owe fully one in five deaths globally to unhealthy diets alone. People are eating too much sugar, refined foods, and processed meat, and this contributes to the diseases of our modern civilization. And it's not just the ingredients; it's the portions. Foods today are often engineered for overconsumption. As previously noted, we become overfed yet undernourished. We'll see how the same can be said for many of our dogs, too. In a study of 3,884 dogs in England, 75.8 percent were diagnosed with one or more health disorders during their first veterinarian "wellness visit."

As we all know, obesity has become a major problem throughout much of the world, especially in developed, high-income countries. We use this word delicately but with good intention. Awareness can bring action. Despite spending trillions of dollars on research and drug development, we know that the risk of developing cancers, cardiovascular disease, and neurodegenerative maladies has continued to increase and is linked to dangerous excess weight. But what about our dogs? They are packing on the pounds, too; more than half of US pets are overweight or obese. There are many contributing factors as to why pets are fat, but knowing that the pet food industry grew to a value of $60 billion in less than sixty years helps to illuminate the largest contributor.

---

* The term "Blue Zones" first appeared in a November 2005 *National Geographic* cover story, "The Secrets of a Long Life" by Dan Buettner. The concept grew out of demographic work done by Gianni Pes and Michel Poulain and outlined in the journal *Experimental Gerontology* the previous year. Pes and Poulain had identified Sardinia's Nuoro Province as the region with the highest concentration of male centenarians. As the two demographers zeroed in on the cluster of villages with the highest longevity, they drew concentric blue circles on the map and began referring to the area inside the circle as the "Blue Zone." Buettner then broadened the term with Pes and Poulain as they found other longevity hot spots in the world, including Ikaria, Greece; Okinawa, Japan; Loma Linda, California; and Nicoya Peninsula, Costa Rica.

Overweight (including obese) dogs have been studied for many years, and the two biggest inputs to their weighty predicament appear to be (1) how and what we're feeding them, and (2) how much exercise they're getting. Interestingly, a 2020 Dutch study of a little more than twenty-three hundred dog owners revealed that "permissive parenting" leads to overweight and obese dogs, just like permissive parenting in the human world relates to overweight (and misbehaving) children. In their study, owners of overweight dogs were more likely to see them as "babies" and to allow them to sleep on the bed, while not prioritizing diet and exercise. These overweight dogs also were more likely to have "a number of undesirable behaviors," including barking, growling, or snapping at strangers, being fearful of the outdoors, and ignoring commands.

Contrary to conventional wisdom, **dogs don't have a carbohydrate requirement, and the average bag of grain-based food is often more than 50 percent carbs, largely from insulin-raising corn or potatoes**. That's like diabetes in a bowl of kibble, with "cides" (i.e., pesticides, herbicides, and fungicides). In addition to corn being rich in carbohydrates, it raises blood sugar levels in dogs quickly and it's heavily sprayed—receiving 30 percent of all agricultural chemicals used in the United States. Grain-free dog food fares no better, at about 40 percent sugars and starches, on average. Don't be fooled by the "grain-free" labeling that seemingly shouts "healthy." Some grain-free kibble has the most starch of any pet food. The labeling practices in the pet food world rival the deceptions we see in our grocery stores, as you'll soon learn. A starch-heavy diet sets the stage for a host of degenerative diseases that can be avoided by choosing less metabolically stressful foods.

We endorse a minimally processed, fresh, whole-food diet (and we'll define exactly what that means), with variety, just as you'd like to eat. **Replacing as little as 10 percent of your dog's daily processed pet food (kibble) with fresh food creates positive changes in a dog's body**, so it's not an all-or-nothing mentality when it comes to improving your dog's health. You can simply change the kinds of treats you give your fur ball to meet that 10 percent shift: Swap out the commercial dog treats you wouldn't dare try yourself for something you'd actually eat, like a handful of blueberries or raw, bite-size carrots. Every small step you take can result in significant overall health benefits. We're going to make this practical, economical, and doable timewise. Once you grasp the power of food, the motivation to change your ways will follow, and we've got plenty of step-by-step guidance later on in the book.

Food is one of the most potent ways to build or destroy health in our companions (and ourselves); it can either heal or harm. And you cannot supplement your way to compensate for poor-quality food—that's like taking a multivitamin while eating junky fast food every day. Chasing your sugary soda addiction with juice cleanses will not save you.

Like most medical students, veterinary students are not taught extensive nutrition in their education. But many veterinarians have evolved their thinking from "It doesn't matter what you feed your dog with cancer as long as he eats" to recognizing that food choices play a major role in immune response and disease recovery. Nutrigenomics, which is the study of how nutrition and genes interconnect in areas such as disease prevention and treatment, is necessary for understanding the health of all dogs. Nutrigenomics offers our dogs the possibility of reversible destiny. We actually met each other over the subject of pet nutrition. Sammie, Rodney's Shepherd, nearly lost her life before her first birthday. Her kidneys had been destroyed by tainted jerky treats promising "to support joints, promote immune health, and benefit the skin coat." A second opinion before she was euthanized saved Sam, who then went on a specialized homemade diet to rescue her kidneys. The experience remains a testament to the power of food as medicine. A few years later, Sam's cancer diagnosis ultimately brought us together as a duo seeking the clues to optimal pet health through diet and the link between pet nutrition and pet longevity. That's when we rolled up our sleeves and got to work. The time had come to excavate all the science hanging out in medical and veterinary journals and share it with the world.

ORIGIN STORIES

I (Rodney) got Sam to soothe dashed dreams during a tumultuous time in my life. You see, as a first-generation Canadian, I grew up in a traditional Lebanese home filled with family and plastic-covered furniture and exactly zero pets. Although I was a poor student, I found success on the football field with Team Canada, and dreamed of one day playing for the Canadian Football League until I blew out my knee. Two life-changing things then happened: I abandoned my football dreams and, while recuperating, watched the movie *I Am Legend*. In the film, Will Smith plays a man struggling to survive in a lonely, postapocalyptic world. His constant companion, protector, and only friend is a German Shepherd named Sam; the two have a deep, vital, symbiotic relationship. As I watched the movie, something in me clicked. Until that point, "human-animal bond" were just words to me. I sensed, however, that there was a whole world of connection I was missing: the possibility of a life-enriching relationship that could exist between man and beast. As my knee healed and my football dreams faded, I did the only logical thing: I got my own German Shepherd. Her name, naturally, was Sam, and her arrival in 2008 changed everything for me.

For me (Dr. Becker), my love of animals dates back as far as I can remember. My parents' first inkling that I was serious about helping animals was on a rainy day in Columbus, Ohio, circa 1973. At age three, I frantically implored my mother to help me rescue the worms I found "stranded" on a sidewalk near our home. (Mom obliged.) From that day forward, my parents nurtured my passion for all animals, although they did give me one firm condition: Any animal I brought into the family home had to fit through the front door. It did not take me long to find my place in the world. By the time I was thirteen, I was volunteering at the local Humane Society; at sixteen, I was a federally licensed wildlife rehabilitator. A few short years later, I was a veterinary student, on my way to turning my passion into a profession. A more proactive, integrative approach to animal care suited my beliefs, interests, and personality. It was common sense to me to start with the least toxic and least invasive medical options; it was even more logical to prevent the body from breaking in the first place. In the following years I became certified in rehabilitation therapy (physiotherapy) and animal acupuncture, wrote a pet cookbook, and eventually founded the first proactive veterinary hospital in the Midwest.

Throughout my career, however, the lessons I learned from my family pets remained top of mind in my professional life. For example, our family dog Sooty lived to be nineteen; he proved to me that lifestyle factors really matter. Because of finances, Sooty ate a base diet of kibble, but the myriad other excellent lifestyle choices made throughout his life clearly made a significant difference. Gemini, a rescue Rottweiler I adopted during my first year of medical school, showed me that food is vitally important—in fact, my homemade meals brought her back from the brink of death. Gemini was my first Forever Dog, far surpassing how long she was predicted to live, in part because of the proactive strategies I began implementing the moment I adopted her. To this day—and even though I've had as many as twenty-eight pets (including lots of amphibians, reptiles, and birds) at once—Gemini remains the patient whose long journey through sickness and health taught me the most.

Pet health goes far beyond what we feed them. There's a lot more to good medicine than food alone. It bears repeating that dogs are exposed to the same pollutants and carcinogens as humans. Importantly, the health choices that allow us to live longer generally serve the same purpose for our dogs.

Two good questions: Are dogs living longer today than their ancestors did? Are they living *better*?

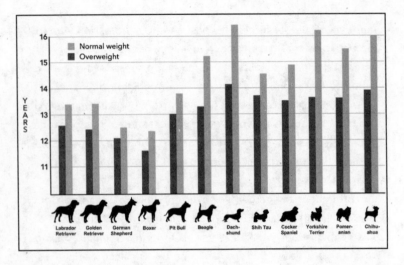

Living an extra year or two because you're a dog of normal weight may not seem like much, but in dog years that's huge. There's no question that dogs have earned longevity gains alongside us. Dogs' life expectancies have increased since their evolution from their ancestral wolves, just as we now outpace our forebears. But that upward trend on the life span may be reversing course—at least when it comes to healthy years. The quality of a dog's life is not what it once was. Although we don't yet have the longitudinal scientific proof that during our lifetime the overall life expectancy of dogs has declined in recent years, there is plenty of anecdotal evidence and a growing body of research that compellingly points to a new and alarming trend. In the UK, for instance, a study of pedigree dogs in 2014 revealed a significant decrease in longevity over the previous decade, with the Staffordshire Bull Terrier losing an average of three full years of life span. In only a decade, Britain's purebred dogs' median lifespan fell by 11 percent. Findings from a study out of the University of California, Davis, suggested that mixed breeds don't necessarily have a lower rate of genetic disorders. Of the 90,000 records reviewed, 27,254 involved dogs with at least one of twenty-four genetic disorders, ranging from different forms of cancer, heart disease, endocrine system dysfunction, orthopedic conditions, allergies, bloat, cataracts, eye lens problems, epilepsy, and liver disease. The study found that thirteen of the twenty-four genetic disorders were similarly pervasive between purebreds and mixed breeds (this debunks the commonly held notion that mutts live longer).

Dogs and humans seem to have hit a proverbial existential wall. Although some experts narrowly blame the change in dog longevity solely on closed gene pools, popular sires, or a preference for aesthetics (looks) over health, the science tells us otherwise: environmental influences, including a lifetime consumption of fast food, and assorted physical, emotional, and chemical stressors play critical—and long-understood—roles in longevity. While many things factor into any human being's risk for premature death, human beings are relatively monolithic, homogeneous creatures; we're fundamentally pretty similar. Dogs, by contrast, come in a wide variety of breeds and sizes; as a result, their health risk profiles are exponentially more complicated to distill and understand. We also can't lose sight of the difference between a good, long, healthy life and a miserable, long, disease-ridden life.

Bramble, a blue Merle Collie from the UK, once held the Guinness World Record for being the oldest living dog at twenty-five years old, which is like a human living long past one hundred! Bramble lived on a healthy, high-quality homemade diet and led a very active, low-stress life. According to her owner, who chronicled the dog's life in a book with the secrets to Bramble's longevity, "Dogs are better being educated rather than trained . . . Learning to communicate with dogs is an important first step." We couldn't agree more; solid relationships are built on trust and excellent *two-way* communication and a mutual understanding of each other (this is true for any relationship!). Which begs the question: How well are we listening to our dogs? She also astutely observes: "Even with the best will in the world and the kindest owners these animals are in our homes at our behest and in this they have no choice."

Unlike the average life span, there is no statistic to mark the end of the average health span. The World Health Organization has developed an indicator, HALE (healthy life expectancy; pronounced *haley*), to address this gap. It factors out disabling illnesses and injuries to compute the actual number of good, "full health" years that a newborn can expect to live. In other words, the calculation aims to tell you how long, on average, a person will live a healthy life before those illnesses and disabilities begin to rob one's quality of life.

Without getting into the details of this complex equation (leave it to the statisticians and demographers), suffice it to say that the last time HALE was calculated in 2015, the figure (a global average for both genders) was 63.1 years—8.3 years less than total life expectancy at birth. In other words, poor health results in a loss of nearly eight years of *healthy* life. Said another way, on average, around the world, we live up to 20 percent of our lives in an unhealthy state. That's a long time. Conversely, think about gaining 20 percent more of a healthy life. Now, in terms of dogs, consider the following: If disease normally kicks in for dogs at around eight, and the average life span for dogs is eleven, then dogs live 27 percent of their lives in an unhealthy state; we'd venture to guess the percentage is closer to thirty when you think about breeds whose life spans should exceed eleven years.

"Modern" veterinary medicine follows the same reactive medical approach taught to medical students treating humans: that companion animal degenerative disease is unavoidable and should be expected by midlife, culminating in diagnoses with poor prognoses as our pets age. Vets learn protocols to prescribe *after* disease is present, but not a single preventive strategy, aside from weight management, was ever taught during my years of veterinary training. The students in my (Dr. Becker's) wellness medicine rotation in vet school designed vaccine protocols for healthy puppies and kittens. There were no curricula or even discussions about how to *prevent* arthritis and muscle atrophy midlife, how to maintain healthy organ systems as pets age, or how to reduce the potential of cognitive decline or cancer before it happens.

Dr. David Sinclair of Harvard, who studies genetics and the biology of aging and has written extensively about the secrets to a long and healthy life, told us he thinks of aging itself as a disease. By thinking of aging in this way, we can work to "cure"—or at least control—aging. In his mind, treating aging may be easier than treating cancer or heart disease. Dr. Sinclair's admirable perspective and ambition have helped spur antiaging research. Aging itself is a natural, inevitable, and beautiful part of life, unless disease absurdly accelerates degeneration—the forty-year-old who has never smoked is suddenly diagnosed with lung cancer, or the five-year-old Boxer dies unexpectedly from a congenital heart defect. Aging is a part of life no matter what kind of animal you are. But aging faster than normal or dying young should not and need not be part of life in the twenty-first century.

The Happiness Test

Based on our anecdotal surveys of dog owners, a top question they would ask their pups if the pups could talk is "Are you happy?" This question is often followed by "How could I make your life happier?" These are great questions that prompt a third: "Does a pet's health reflect its human's health status?"

In our experience, pets' health complaints often mirror those of their humans.

If your dog is anxious, are you? If your dog is overweight and out of shape, are you? If your dog has allergies, do you? Our pets' health can be an indicator for our own. Anxiety, obesity, allergies, gastrointestinal infections, and even insomnia are all examples of the disorders shared between pet and owner.

Studies on pet-human pairs are relatively new in the research space, but the existing research has highlighted interesting initial findings. In the Netherlands, researchers found that overweight dogs were more likely to have overweight owners. (This shouldn't be a surprise; we see the same thing with children and their parents.) In addition, the more time a human-pet pair spend walking, the less likely they are to be overweight. A separate study from Germany found that owners typically pass on their snacking behaviors and beliefs regarding portions and processed foods to their pets, which in turn determines a pet's daily caloric intake.

In Finland, a noteworthy 2018 study took one diagnosis in particular—allergies—and sought to find a pattern among dog-human pairs. Lo and behold, they found that people and their dogs who live in an urban environment and who are disconnected from nature and other animals have a higher risk for allergies than those who live on a farm or in a household with many animals and children, or who regularly stroll through a forest. Allergies in dogs are often diagnosed as canine atopic dermatitis, which is similar to human eczema, and are one of the most common reasons dogs go to the vet. A Finnish study led by some of the same researchers documented another significant risk factor for canine allergies: eating an ultra-processed carbohydrate-based diet. In their 2020 *PLOS One* paper, the group concluded that feeding a nonprocessed, fresh, meat-based diet early in life protects against canine atopic dermatitis, whereas an ultra-processed carb-loaded diet is a risk factor. They also identified other important variables associated with a significantly lower risk for canine atopic dermatitis: "maternal deworming during pregnancy, sunlight exposure during early postnatal period, normal body condition score during the early postnatal period, the puppy being born within the same family that it would stay in, and spending time on a dirt or grass surface from 2 to 6 months." **Bottom line: Fewer processed carbs and more dirt are key**.

This phenomenon—protection from allergies through exposure to an agrarian lifestyle and its inherent dirt—is sometimes called the "farm effect." Indeed, it pays to get dirty sometimes. The dirt, to be sure, is much more than what we find under our feet out in nature. The star player in the dirt is the community of microbes found in rural and natural environments that serves an important role in protecting against pathogens, supporting metabolism, and educating the immune system so it does not develop hypersensitivities to allergens. The dirt teaches the body how to distinguish between friend and potential foe. Thankfully, there are all sorts of emerging research projects, including the Canine Healthy Soil Project, focused on testing the biodiversity hypothesis that a puppy's early exposure to healthy soil microorganisms can be profoundly beneficial in reestablishing a dog's ancestral microbial communities in and on their body to enhance overall health span for years to come.

In later chapters, we'll be going into deeper detail about this phenomenon, because it's stirring a revolution in scientific circles as we learn more about how the friendly microbes (and their metabolites, or substances formed by their own metabolism) around us contribute to our physiology and health—including that of our dogs. Immunologists the world over are racing to decode the secrets to the microbiome—the aggregate of all microbiota (dominated by bacteria) that live in and on us in a mostly symbiotic relationship. These commensal organisms have contributed to our survival for millions of years and evolved with us.

All of us, dogs included, have unique microbiomes residing throughout the body's tissues and biofluids; they are everywhere, inhabiting our guts, mouths, sexual organs and fluids, lungs, eyes, ears, skin, and so on. The body's ecosystem—human or canine—is more microbial than anything else. It's not surprising that researchers have already documented vast differences in the microbiomes of dogs and humans who suffer from allergies compared to their healthy, allergy-free counterparts. They also have documented significant differences in the intestinal microbiome of healthy dogs who suffer from chronic or acute intestinal inflammation; **there is a strong relationship between the health of a dog's microbiome and risk for gastrointestinal diseases.** Some studies are even beginning to show relationships between our microbiomes and those of our dogs who live with us. In 2020, for example, yet another group of Finnish scientists (including some of the same scientists in the previously mentioned studies) found that dogs and their owners were more likely to suffer simultaneously from allergic conditions in an urban environment and when their exposure to beneficial environmental microbes was limited. Interestingly, they also found that the skin's microbiome, which powerfully plays into skin health, tends to be shaped by the living environment *in both species*. As we'll also see later on, this microbiome develops and thrives from a collection of inputs, ranging from environmental exposures to dietary choices. What you and your dog eat factors mightily into the strength, function, and evolution of both of your microbiomes. In turn, the microbiome impacts risk for illness and disorders from the inside out.

As published in *Nature*, one of the most prestigious science journals in the world, our pets' emotional states, also influenced by the microbiome, can remarkably mirror our own. Anyone who owns a dog knows that dogs and humans read each other very well, and this ability seems to be related to the long period of association between both social mammals during the domestication process. These shared emotions function as "social glue" and help in the development and maintenance of strong and enduring social bonds. When we interviewed lead researcher Dr. Lina Roth about her findings, published in a 2019 *Nature* article, she referred to hair cortisol levels (an indicator of chronic stress) in dog-human pairs and indicated that a strong degree of "interspecies synchronization" exists between the two. A caveat to this "emotional contagion" is that it appears to only travel from humans to pets and not the other way around. Such a discovery lends credence to the deleterious effects our own stress can have on our dogs, often unbeknownst to us. If dogs can understand our emotional and mental states, what does it mean when we're living under the chronic stress of serious trauma or extreme anxiety? Our dogs can be suffering immensely and incessantly, right alongside us. In one particularly alarming study by a global team of researchers, people who have a propensity to run away from their feelings (what's called an "avoidant attachment style") were found to be more likely to have dogs that try to run away and emotionally distance themselves from their owners when faced with a social stressor.

At the University of Naples Federico II in Italy, we watched dogs discern and react within *one second* to human sweat samples collected during happy and fearful emotional states. Dr. Biagio D'Aniello told us the most eye-opening aspect of his research wasn't that dogs could differentiate human emotions via chemoreceptors in their nose; it was that the dogs' own biochemical markers were, in turn, affected. Dogs and humans are emotionally intertwined, and their emotional states impact each other's physiology. The heated argument at the office that raised your blood pressure also changed your hormonal chemistry, with detectable traces of residual stress hormones literally oozing from your pores; your dog will identify (and respond to) those stress hormones when you get home. Ever notice how your dog sniffs you when you return from somewhere? Turns out she's smelling how your day went, finding out if you're okay.

When we asked Dr. D'Aniello how we can help dogs cope with our own chaotic lives, his answer made us pause: "When you get home from work, take a shower. Immediately." He said this with a subtle smirk. A more practical approach, he suggested, is cultivating stress-reducing habits and tools you can implement on a daily basis. **It turns out self-care, including exercise, yoga, meditation, or whatever helps you authentically decompress and return to a state of balanced homeostasis, is a gift to your body, mind, soul . . . and dog.**

Dogs, like humans, are social animals. As you're about to find out, dogs cozied up to us long ago in a beautifully choreographed dance to navigate this planet together and enjoy life as much as possible. It's no wonder our relationships with our dogs are shaped by so many aspects of our lives, from our habits and stress levels to our microbiome. The story of our co-evolution with dogs makes us smile because it's so heartwarming. We can picture Reggie, Sam, Gemini, and all the other dogs who've gone to heaven smiling with us, too.

LONGEVITY JUNKIE TAKEAWAYS

➤ Regardless of your dog's breed and underlying genetics, the health-span goal is the same: to live as long as possible with a high quality of life. That is, by definition, a Forever Dog.

➤ A massive amount of research demonstrates our ability to positively influence and control gene expression by changing our dog's environment. This is called epigenetics.

➤ One of the most powerful sources of longevity medicine is food. Swap out as little as 10 percent of your dog's daily processed pet food (kibble and treats) with fresher alternatives to create positive changes in your dog's body. A great place to start is by switching to unprocessed, fresher treats.

➤ The same factors that have compromised our longevity have also caused our dogs to hit an existential health wall: lack of a diverse, minimally processed diet; overeating; sedentariness (lack of exercise); and environmental exposures to bad chemical toxins and chronic stress.

➤ Our dogs pick up on our stress; engaging in healthy activities and practices that reduce stress and cultivate emotional well-being in our own lives will positively affect our dogs.

2

Our Co-Evolution with Dogs

From Wild Wolves to Willful Pets

To his dog, every man is Napoleon; hence
the constant popularity of dogs.

—Aldous Huxley

In the picture above, we see the skeletal remains of a woman who lies in a fetal position, with her hand cradling the head of a puppy in a gesture of affection. It was discovered in the late 1970s in a twelve-thousand-year-old burial pit by the shores of the Hula Lake, about sixteen miles north of the Sea of Galilee, where a small hunter-gatherer community once lived. Capturing a period when humans made simple stone tools and lived in semipermanent dugout homes with stone walls and thatched roofs, this photograph underscores the deep interconnection of human and canine from the earliest recorded times.

More recently, in 2016 archaeologists unearthed twenty-six-thousand-year-old canine paw prints alongside those of a human child who was eight to ten years old and about four and a half feet tall. The discovery took place inside Chauvet Cave, a Paleolithic site in southern France. It has been hypothesized that the barefoot child was walking, not running, although at one point it appears that he or she slipped in the soft clay. We also know that the child carried a torch because there is evidence of him or her stopping at one point to clean the torch, leaving behind a stain of charcoal. It's astonishing to think of a Paleolithic kid, with a pet dog in tow, exploring this ancient cave, home to some of the world's oldest paintings. Over four hundred images of animals were created around thirty-two thousand years ago (our caveman days).

This finding shatters the established notion that dogs were domesticated only between 12,500 and 15,000 years ago. More important, the new time period radically alters the answer to how dogs became humanity's best friend. There are some theories that humans and dogs lived among each other as much as 130,000 years ago, which is long before our ancestors created communities based on agricultural practices. But this remains a heated debate, and future research will have to suss out the answer. (Even as we write these words, a new headline reads, "Too Much Meat During Ice Age Winters Gave Rise to Dogs"; thus, the debate about the origin of domesticated dogs continues.) It's hard even to define the word "domestication" to begin with, or to determine whether the phenomenon happened once, twice, or multiple times throughout Asia and Europe. Regardless of which theory (or combination thereof) is correct, the indisputable fact remains: "The domestication of dogs has increased the success of both species to the point that dogs are now the most numerous carnivore on the planet." That statement came from a 2021 paper by Finnish researchers for the journal *Nature*.

It's also worth noting that the story of our evolution remains a bit of a mystery, too, with new evidence emerging to show that our timeline and trek across the world might not be perfectly documented (yet). Interestingly, our own DNA does not always reveal parts of prehistory that we can see with dog genomes. According to Pontus Skoglund, a population geneticist at the Francis Crick Institute in London who co-led a study published in 2020 on canine evolution, "Dogs are a separate tracer dye for human history." Indeed, we may have to mine the dog genome more to unveil the details to our own past and migration throughout the world.

No matter what breed of dog we're talking about—Labradoodles, Great Danes, or Chihuahuas—they all have one thing in common: the gray wolf, *Canis lupus.*

While it may be hard to see much resemblance among these very different-looking dogs (other than their common fur, four legs, and barking), every single dog on the planet—all of the more than four hundred recognized breeds—traces back to an extinct wolf species shared with the gray wolf. We can prove it through genetic studies, though to be clear, dogs are not wolves. And our partnership with our pooches is just as ancient. Dogs were the first species to forge a deep bond with humans. (Humans kept animals as companions long before we domesticated livestock like sheep, goats, and cattle some ten thousand years ago. By contrast, horses were tamed in Eurasia only around six thousand years ago. Though they weren't household pets, horses inspired passionate feelings in their owners.)

One of the first pet dogs whose name is known to us is Abutiu (also transcribed as Abuwtiyuw), who belonged to an Egyptian pharaoh in the early third millennium BC. It is thought that he was a Sight Hound, a lightly built hunting dog similar to a Greyhound, with erect ears and a curly tail. After Abutiu's death, his heartbroken owner gave him a royal burial. The inscription on the block of limestone tomb explained, "His Majesty did this for him in order that he might be honored before the great god Anubis."

The origins of selective breeding also remain up for grabs scientifically, especially the roots of some particular varieties of canines. An example of this occurred in a study of herding dogs conducted by the National Institutes of Health. When researchers looked at herding dogs from different regions (the United Kingdom, north-ern Europe, and southern Europe), they hypothesized that the genetic makeup of these breeds would be similar. Their findings, published in 2017, showed otherwise. In fact, researchers found that the way each group herded their flocks differed depending on their genetic makeup. This result fuels the idea that humans bred dogs with a purpose in mind.

A majority of the dog breeds we know today were created within the last 150 years in an era called the Victorian explosion. During this time period, Great Britain took dog breeding and turned it into a scientific hobby and sport. The end result is four hundred–plus types of dogs recognized today as distinct breeds (note: not all modern breeds came out of the UK, as the trend in breeding dogs swept the world). This shift in favor of a desired canine aesthetic has been accompanied by devastating health consequences. This was at the height of Darwin's game-changing work and writings; he himself was obsessed with dog breeding and befriended top dog fanciers. But if you scroll through pictures of dog breeds from the nineteenth century and compare them to their current counter-parts, you can see that dramatic changes have occurred. Rigorous selective breeding for specific physical traits throughout the twen-tieth century has given us Dachshunds with shorter legs, German Shepherds with stockier builds and sloped backs, and Bulldogs with more pronounced facial wrinkles and a thicker, squatter body (in fact, few dogs have been as artificially shaped by breeding as the English Bulldog). Such a shift has not come without downsides and certain forfeitures in the health realm. There's been a dual impact: a tremendous loss of genetic diversity and an acquisition of unde-sirable heritable diseases.

Many people think "mutts" are healthier than purebred dogs, but this isn't always the case, as we've mentioned. According to veterinary epidemiologist Dr. Brenda Bonnett, who is now CEO of International Partnership for Dogs: "Many genetic diseases are the result of ancient mutations and are distributed widely in all dogs. Some genetic diseases have increased in frequency with inbreeding to create breeds, but may be different diseases at different levels in different breeds." She offers a good case in point: If crossbreeding is done using the absolutely healthiest, disease-free, and genetically robust Poodle and the most equally perfect Labrador, chances are that the offspring stand a good chance of being healthy (though there are never guarantees). However, mating just any two dogs cannot be presumed to produce mixed-breed puppies that are healthier (hence, the issues with puppy mills and backyard breeders producing "designer mutts" with no health clearances). This is especially true for those ubiquitous, ancient disease mutations. When we asked her about any potential value to DNA testing dogs so pet owners can identify what genetic diseases might be lurking and to watch out for, she said that **while genetic testing can be beneficial, many of the most common and important diseases in dogs are not detected by genetic testing—*yet*.**

Genetic testing is rapidly moving into the canine world. In North America, Embark and Wisdom DNA tests are currently the most popular breed identification kits owners are buying to learn more about their dogs' breeds, ancestry, and disease markers. The upside to these tests is that you *can* screen for more than 190 heritable diseases that have been identified in dogs via specific genetic disease markers, so it's imperative for excellent breeders to use this testing to continue to ensure well-bred, healthy dogs. These tests help discern mass-produced purebred dogs (coming from factory farms and puppy mills) from well-bred dogs coming from preservation or functional breeders committed to improving canine health. Not all purebred dogs are well-bred dogs, so if you're going to spend money on a puppy, it's imperative you do your homework and partner with a breeder who is committed to improving health span through thoughtful, well-designed genetic pairings. This critical step didn't happen for Reggie and Sam: Their breeders produced litters without determining their genetic well-being and compatibility, bringing about adorable fur balls that didn't live as long as they deserved.

If a pet parent does genetic testing on a mixed-breed rescue (a mutt) or a pet store (puppy mill) dog, it's really important to remember that even though the dog may test positive for carrying genetic variants for diseases, it doesn't mean the dog will automatically express those genes and get the disease. Every vet we know can tell horror stories about clients who discover their dogs are carrying some "bad DNA" and make crazy decisions, based on nothing other than a piece of paper. Vets are often hesitant about pet owners testing for genetic diseases, because the results do not tell us anything about whether the dog will actually get the disease.

Some people tell us they don't want to know, so they don't test. Other people do the testing, discover their dogs are carriers for heritable diseases, but forget that these genes may never be expressed; they then spend their lives feeling anxious about something that may never come to fruition. If you've done genetic testing that shows your dog has variants or is a carrier for known health risks, don't panic. Ideally, if you discover your dog has some predisposing DNA alterations, you will view this discovery as an early-bird opportunity to preemptively initiate a therapeutic nutrition and lifestyle plan with the goal of positively modulating your dog's epigenetics.

Our DNA controls virtually everything about us, so identifying the genetic disease markers in our bodies empowers us to take proactive steps with our lifestyle, but that's just the beginning of how getting to know our own DNA can improve our health and life span. The reality is we all carry some less-than-lovely DNA, but that's where epigenetics (more on this shortly) and nutrigenomics come into play. **Just like with you, the food your dog eats, the carcinogens he's exposed to, and the lifestyle you create for her can increase or decrease the likelihood that heritable diseases will express themselves.** We wrote this book to help you recognize and mitigate lifestyle health obstacles, in order to maximize your dog's health span and, ultimately, life span, regardless of the genetic hand he or she was dealt.

We will explore why our furry friends may be diagnosed with the same conditions as humans, including cancer, heart disease, and obesity. We needn't look further than what's happened in our own evolution. Since our caveman days, we've gotten good at making life easier, but at certain costs. We will address those costs up front, but don't worry, we also will directly address how to counteract them in Parts II and III.

Mass Migration and Agriculture

Raise your hand if you've ever tried a particular diet—low-fat, paleo, keto, vegan, carnivore, pescatarian, you name it—to either lose weight, manage or conquer a condition, or just try to be a healthier person in general. We've both had our thrills with various diets and today enjoy a mostly meat-free lifestyle with intermittent fasting. **Diets are cornerstones of disease and, conversely, health**. That old saying is true: we are what we eat. But what about our dogs? What's best for them? Is feeding them the exact same bowl of kibble, daily, ideal? (Think about that for a moment: Would you like to eat the same exact thing at every hunger pang? Although we are told this is anthropomorphizing and dogs don't care, offer your dog three bowls containing three different foods and watch: he won't go back to the same bowl over and over. Even Pavlov's dogs needed variety.) We'll get to the details in Part II, but here we're laying the groundwork for that conversation by taking a tour of how our sustenance through the ages has changed—and shaped—us, especially with respect to the development of agriculture.

Around twelve thousand years ago, what we'd call "disruptive technology" took root, literally. We began to mobilize, organize, and settle into communities as we moved toward an agriculture-based lifestyle and abandoned the hunter-gatherer way of life. This shift was responsible for an increase in population growth but a decrease in the quality of our diets. Humans began to eat more calories than necessary after discovering how to farm, grow, and store crops, especially grains like corn and wheat; variety in the human diet also suffered, with a focus on fewer types of foods. Scholars who study the effects agriculture had on human society note that while it had its merits, it also came with consequences that only deepened the more advanced and sophisticated we got with our husbandry. Later, farmed wheat and corn were turned into highly processed fare—the white breads, hot dogs, and junk foods of the world—and exposed us to the chemicals, some of which are proven carcinogens, that are used in modern farming.

As a UCLA professor, a Pulitzer Prize-winning author (*Guns, Germs, and Steel*), and a leading historian, anthropologist, and geographer, Jared Diamond has contributed a vast amount of literature about the impact agriculture has on human health. Over the years he's made bold statements, going so far as to call agriculture "the worst mistake in the history of the human race." He writes that hunter-gatherers' diets were more diverse than that of early farmers, whose main source of energy came from a small variety of carbohydrate-based crops. Diamond also affirms the belief that trade spawned by this agricultural explosion may have contributed to an increase in parasites and infectious diseases. He has stated that the adoption of agriculture "was in many ways a catastrophe from which we have never recovered." Another historian, Yuval Noah Harari, added his voice to the idea in his bestselling book *Sapiens*: "The Agricultural Revolution certainly enlarged the sum total of food at the disposal of humankind, but the extra food did not translate into a better diet or more leisure . . . The Agricultural Revolution was history's biggest fraud." You may not agree with these historians, but one fact remains clear: The agricultural revolution substantially impacted man's best friend.

When we choose what to eat, we choose what information to give our bodies. That's right: **Food is information for our cells and tissues all the way down to their molecular structure**. This is true whether we're talking about a human being, bumblebee, birch tree, or Beagle. Dr. David Sinclair agrees, telling us that one cause of aging is "a loss of information in the body."

If you have not thought about food in this context before, consider the following: Food is way more than energy. As we digest, its nutrients facilitate communication between the environment and our DNA. Those signals have the power to *influence how our genes behave* and how our DNA is turned into messages that affect the functionality of our bodies. This translates to the ability to change the behavior of your DNA. These alterations caused by extrinsic influences involve a field of study called "epigenetics." The great news is that we have an active role in affecting which genetic switches get turned on or off. Let us give you one quick example that holds true for both humans and dogs: A pro-inflammatory diet heavy in refined carbohydrates negatively impacts a gene important for brain health known as "brain-derived neurotrophic factor" (BDNF). This gene codes for a protein also referred to as BDNF that is responsible for the growth and nourishment of brain cells. We like to think of BDNF as fertilizer for the brain. BDNF doesn't come in supplement form, and you won't get it from food, but there are things we can do to keep our dogs' bodies producing it as they age. And the right food can support the body's own BDNF-producing powers. When we eat healthful fats and proteins, a style of eating that was common among our preagricultural ancestors (and their canine companions), the gene pathway fires up, increasing BDNF production. In essence, you support your brain's health. Exercise also increases the production of BDNF. Stress levels and sleep factor into BDNF production as well; in fact, reduced levels of BDNF are now associated with insomnia, and studies reveal that it can be a vicious cycle whereby high stress levels dampen BDNF production, which in turn disrupts sound sleep. Studies further show that people who suffer from cognitive decline and neurodegenerative diseases have low BDNF, and those who maintain high levels continue to improve their learning and memory while staving off brain disease.

Dogs have been excellent models for showing the effects that certain lifestyle habits can have on boosting BDNF and cognitive performance. A 2012 study done at McMaster University in Ontario, Canada, demonstrated that old dogs could physically turn the clocks back on their brains through a combination of "environmental enrichment" and an antioxidant-fortified diet. The environmental enrichment protocol entailed regularly socializing dogs, exercising them, and putting demands on their cognition so they had to think and perform tasks. The researchers documented measurable increases in BDNF in these old dogs that approached levels found in the young dog brain. In other words, simple lifestyle strategies de-aged the dogs.

CARBOHYDRATES CAN BE SPLIT INTO THREE MAIN CATEGORIES:

Sugars: glucose, fructose, galactose, and sucrose (dogs can make glucose from protein in a process called "gluconeogenesis," so supplying sugar in the diet is unnecessary)

Starches: chains of glucose molecules that turn into sugar in the digestive system

Fiber: roughage that our dogs can't absorb but is required for their gut bacteria to create a healthy microbiome

Carbs come from plants (e.g., grains, fruits, herbs, and vegetables) that contain varying amounts of sugar (called a "glycemic index") and different types of fiber (critical for building and fueling the gut microbiome), as well as other health-conferring phytochemicals that can be passed up the food chain. Dogs need fiber and phytochemicals to achieve their maximal life and health span. They don't need large amounts of sugar or starches. **The goal is to feed low-glycemic, fiber-rich "good carbs" that nourish your dog's gut and immune system, and avoid feeding high-glycemic, refined "bad carbs" that provide excessive sugar and create metabolic stress.** We'll show you how to calculate the level of "bad carbs" (aka sugar) in your dog's food in Chapter 9, as this is one way we assess the long-term metabolic stress of your dog's food.

The idea that our DNA works best with an ancient, or ancestral, diet has been the underlying force behind popular diets that aim to minimize carbs, especially the processed ones, and maximize healthful fats and proteins from wholesome sources. Humans and dogs have eaten a more diverse diet containing healthy fat and fiber and fewer refined carbohydrates for more than 99 percent of our existence. We also have evolved to eat far less frequently than the typical person today. Most of us—and many of our dogs— have access to food whenever we want; we love our snacks, treats, and twenty-four-hour drive-throughs and delivery apps that can drop provisions at our doorstep within minutes of a finger swipe. But this Western diet intended to make our lives easier limits our DNA's ability to protect health and longer lifespans. And despite amazing, twenty-first century advances, we still feel the ramifications of this disconnect. So do our companions. When the agricultural revolution happened, we shared grains with our dogs and actually changed their genome. We know from science that dogs produce more pancreatic amylase (the enzyme that breaks down carbs) than wolves.

If you thought agriculture changed the trajectory of our existence, consider what the next phase—Big Ag—has done to us. Big Ag refers to corporate farming that often leads to a bevy of ultra-processed foods. To be clear, ultra-processed foods are not modified

foods. According to a Brazilian group of nutritionists and epide-
miologists at the University of São Paulo, they are best defined as
"formulations mostly of cheap industrial sources of dietary energy
and nutrients plus additives, using a series of processes (hence
'ultra-processed'). All together, they are high in unhealthy types of
fat, refined starches, free sugars and salt, and poor sources of protein, dietary fiber, and micronutrients. Ultra-processed products
are made to be hyper-palatable and attractive, with long shelf-life,
and able to be consumed anywhere, any time. Their formulation,
presentation and marketing often promote overconsumption."
We'll dive into how to determine the processing level of your dog's
food in Part III.

While we moved toward a diet rich in processed products, dogs
also increasingly got their fill from highly processed grub. In the
last century, dogs in modern societies were weaned onto entirely
processed food. Few dogs today have access to unprocessed, whole
foods or minimally processed sustenance. Veterinary students are
taught that this is ideal for companion animals and food production
animals alike; feeding formulated, pelletized, and fortified diets to
factory-farmed animals (concentrated animal feeding operations or
CAFOs) and pets their entire lives has become norm. Many animals (including kids) no longer eat meals made from identifiable
real, whole foods: It's mostly refined, remixed, and repackaged into
bite-size balls of stuff we hope contains enough nutrients to prevent
disease.

Processed foods are items that get infused with ingredients like
sugar (starch), fat, and salt to help keep them edible longer.
Ultra-processed foods are often made in factories, broken down
from their whole or fresh form, and treated with thickeners, colors, glazes, palatants (ingredients that make the food addictive,
from the word "palatable"), and additives to extend shelf life.
For humans, they may be fried before they're packed in cans or

wrappers. For pets, they're extruded: meaning they're cooked under pressure at high temperatures to create crunch. They both might contain protein isolates or interesterified oils (engineered replacements for trans fats, which are now widely banned) or, in the case of many brands of kibble, sprayed with used restaurant grease. You'd be surprised by how similar Snausages and Cheetos are in terms of nutrient value!

Study after study shows that junk foods aren't just bad for our health, they cause people to eat more and gain weight, without providing any additional vitamins and minerals. They are linked to higher rates of cancer and early death. A lot of us have gotten the memo but haven't relayed that information to our dogs.

Vets who encourage people to feed dogs "pet food" their whole lives argue that pet foods have been designed to meet nutritional needs, while junk food clearly does *not* meet nutritional needs. There are a few ultra-processed human foods that are labeled as "all-in-one" nutritionally complete foods: Total brand cereal claims to meet 100 percent of the recommended daily allowances of vitamins and minerals for people, as do some liquid beverages such as Ensure and Soylent. These are the best comparisons to what we've been told to feed our pets for a lifetime: scientifically formulated "everything-you-need" pellets. Granted, many humans consume all-in-one drinks at the beginning and end of their lives, and sometimes in between, when we're on the go or during hospital stays. But nutritionists have never recommended using these "nutritionally complete" products as a sole source of lifetime nutrition. Even "scientifically formulated" infant formula, which nourishes millions of babies around the world every year, is replaced after several months with real foods that are less processed and more diverse. The only family members eating ultra-processed foods *their whole lives* are our pets.

Some people don't see commercial dog food as "processed" to the same degree as processed human food. But by many definitions, **dog food is even *more* processed than any human food**, as you'll learn in Chapter 9. If you knew how kibble was made, you'd begin to see the difference. There are new, highly processed food products hitting the human and pet food industries weekly. These convenience foods are made with a long list of ingredients that already have a long processing story to tell, prior to ending up in a snack food recipe. All of these highly altered bulk ingredients bear no resemblance to any historical relationship they may have had with agriculture crops or commodities. Likewise, there's nothing even remotely close to "fresh" ingredients in any ultra-processed pet food; the bulk ingredients used in commercial pet foods have already been extensively processed (e.g., meat and bone meal, tallow, corn gluten meal, rice bran, etc.) before reaching their final resting place in dry pet food, not to mention the fact that the finished product is expected to be shelf stable for more than one year under ambient (room temperature) conditions (companies have not released research on how long it's safe to feed dogs pet food once the bag has been opened). There's nothing fresh about any aspect of food being extensively processed, for humans or pets.

Pet food processing also impacts the quality of the vitamins delivered, with many vitamins lost in production. Adding insult to this unhealthy picture is the fact that crop residues, including glyphosate, turn up in commercial pet food. Glyphosate is the main ingredient in the lawn chemical Roundup and is most likely a carcinogen. Unfortunately, its widespread use in conventional farming makes it relatively easy for it to land in foodstuffs meant for the commercial dog market. In an alarming 2018 study, Cornell University researchers found detectable glyphosate in *all* eighteen commercial dog and cat food products they tested (including the one GMO-free product in the group) and concluded that "exposure to glyphosate via food consumption is likely higher for pets than humans." They calculated that our pets' exposure to this likely cancer-causing agent is between four and twelve times higher than it is for us on a per kilogram basis.

Another tagalong contaminant found in many commercial dry dog foods is mycotoxins. Mycotoxins are toxic chemical substances naturally produced by fungi that infect many grains, including those in pet foods, and are a common reason for recalls. The December 2020 US recall of a kibble contained enough aflatoxins, one type of mycotoxin, to kill more than seventy dogs and make hundreds more very sick. Mycotoxins create havoc in your dog's body, from organ disease to immunosuppression to cancer, and their effects are well documented. Pet food companies are not required to test finished pet foods for mycotoxin levels. In one US study, nine of the twelve dog foods tested were positive for at least one mycotoxin, results that were consistent with findings from Austria, Italy, and Brazil. If you're feeding your dog kibble made with grains, you're undoubtedly feeding her mycotoxins; the only questions are how much and what's the impact. Don't panic, we'll give you mycotoxin-mitigating strategies.

Logic makes us question the nutritional standards the modern pet food industry has promoted, despite the lack of lifetime studies looking at dogs who eat ultra-processed diets versus those that live on "fresher" foods. In the United States, it's estimated **humans consume about 50 percent of their daily calories via ultra-processed foods; for many pets, at least 85 percent of their calories come from ultra-processed foods.**

We repeat for effect: Our pets don't get to choose what they eat. Like young kids, they eat the foods we put before them, and we commonly overfeed and undernourish them, leading to a myriad of potential health and behavior problems. Human nutritionists recommend consuming less-processed foods, but most veterinarians still recommend *only* processed foods. Where's the disconnect?

To drive home the powerful difference between a processed diet versus a largely fresh or minimally processed one, we'll give you a great example of a recent study that proves the point: In 2019, a research group from Switzerland and Singapore put two diets to the test using healthy Beagles. After leveling the playing field by putting all sixteen dogs on the same commercial dry diet for three months, they measured the dogs' blood fats to record a baseline. Then they randomly split the dogs up into two groups. One set of dogs was fed a commercial diet similar to the first one for three months, and the other group got a homemade, nutritionally complete diet supplemented with linseed oil (flaxseed oil) and salmon oil. Which set showed a better blood fat profile after the experiment? The Beagles getting the omega-3s from healthy oils added to their fresh fare rather than processed dog food were absorbing more of those good fats.

Their blood work was by far richer in omega-3s and lower in unsaturated and monounsaturated fats compared with the commercially fed dogs. Studies like this show that the actual composition and source of food can make a difference in health outcomes, which is especially important for dogs with recurrent skin and ear issues rooted in immune system imbalances. Again, another win for real food over processed.

Boxed In

In 1900, the ratio of rural inhabitants to city dwellers was seven to one. Today, half the globe's population lives in an urban center, and it has been estimated that by 2050, almost 70 percent of us will live in cities. And we spend more than 90 percent of our time indoors. We lack the need to physically move and hunt for food just to survive. Pretty much everything we require is a click or swipe away. In the modern world, we are surrounded by walls, in rooms with artificial lighting and a controlled atmosphere that can deceive our natural circadian rhythms and prevent us from engaging in the kinds of activities our bodies—and our DNA—expect and demand. We interact with nature through windows, our phones and computers, and—less often than we should—a walk outdoors. This means our dogs are increasingly living in the confines of buildings with limited access to nature (except for those coveted walks). They may even miss out on natural sunlight throughout the day if they are left at home with the curtains drawn. We know of countless small dogs whose owners confess that their pup has never touched the earth; they've never luxuriated in a patch of lush grass, felt the force of a strong breeze, pooped in the dirt, or sniffed the fall foliage. That may sound absurd, but it's possible, if you live in a high-rise city building with no backyard and surrounded by concrete sidewalks.

We also know what the science says: Compared to country dogs, city dogs that live indoors the majority of their lives are more likely to have higher levels of anxiety, exhibit signs of more biological stress in their blood work (i.e., higher markers of inflammation and oxidative stress), lack proper exercise, and even suffer from social dysfunction (because they never get to play freely with their canine companions and other people). Dogs are increasingly isolated—caged by their home, a small room or crate, some completely litter-boxed trained, and always leashed when outside. Rather than patrolling expansive rural properties or even a backyard, they are constricted to small spaces and deprived of the wilds of nature. We should add that they can be more vulnerable to certain exposures we ignore. For example, dogs can more easily sense electromagnetic fields (EMF), making it difficult for them to navigate our highly networked and increasingly connected world. Without sounding too much like conspiracy theorists, we know that many dogs dislike being near 5G routers. Shouldn't we notice and protect their sensory experience by keeping them away from these devices? Beyond magnetism, and their amazing ability to use the Earth's magnetic fields to find their way home, they have other super senses in their ears and nose that exceed our own in many ways. Years ago, my childhood dog Sooty Becker escaped through an open garage door during the chaos of moving into a new house. The next morning we found him at the front door of the house he'd grown up in, more than ten miles away. And ten miles is nothing.

You may have heard about Bucky, the black Lab that walked over five hundred miles after his dad moved from Virginia to South Carolina? Apparently he preferred his old stomping grounds. Dogs have a "sixth sense" that may be related to magnetotactic bacteria in their gut; these are bacteria that orient themselves along the magnetic field lines of the Earth's magnetic field. This makes us wonder about all the poor dogs with gut issues and dysbiosis, not to mention all the chemicals dogs inadvertently eat and inhale. Dysbiosis (literally, "bad mode of life") is a condition whereby there's an unhealthy imbalance among the types of organisms present in the body's natural microflora, specifically in the gut. Bearing in mind the acute sense of smell (and even a remarkable ability to detect heat from a distance using their nose) and hearing like radar, we have to raise the question: What kind of urban environmental offenses are harming our animals? Because this is the reality of life for many of us and our dogs, how can we better manage these metaphorical barbs?

We are only just beginning to understand what living in a "built environment" is doing to us—and our furry friends. (When we say "built environment," we mean the human-made areas in which we work and live, including parks, buildings, bridges, roads, and more.) In 2014, the Mayo Clinic, in collaboration with the wellness company Delos, launched a major project called the Well Living Lab to gather facts on how buildings (and what's in them) affect our health and well-being. Scientists already have documented staggering associations. For example, children born today—or anytime during the industrial age—are more at risk of developing diseases like asthma, autoimmune disorders, or food allergies than in prior times. This could be due to what the "hygiene" or "microbiome hypothesis" theorizes: the fact that most humans now interact less with the microbes found in nature. This helps explain why studies on allergies in dogs (notably canine atopic dermatitis) also show a correlation between living in an extremely clean household and an increased risk for allergies.

The Environmental Working Group (EWG) was among the first organizations to look into the extent to which pets are exposed to contaminants in our homes and outdoor environments. And what it found was surprising: The same industrial chemicals found in people exist in *higher* numbers in pets. This finding makes clear that American pets are acting as patient zero for the chemically induced roots of many of the health conditions humans and animals face today.

In 2008, the EWG performed a seminal study of plastics and food-packaging chemicals, heavy metals, fire retardants, and stain-proofing chemicals in pooled samples of blood and urine from twenty dogs and thirty-seven cats collected at a Virginia veterinary clinic. It found that dogs and cats were contaminated with forty-eight of seventy industrial chemicals tested, including forty-three chemicals at levels higher than those typically found in people. Average levels of many chemicals were substantially higher in pets than is typical for people, with 2.4 times higher levels of stain- and grease-proof coatings (perfluorochemicals) in dogs, 23 times more fire retardants (PBDEs) in cats, and more than 5 times the amounts of mercury, compared to average levels in people found in national studies conducted by the Centers for Disease Control and Prevention (CDC) and the EWG. Perfluorochemicals (PFCs) are more pervasive than most people think. They are used in surface coating and protectant formulations for paper and cardboard packaging products; carpets; leather products; and textiles that repel water, grease, and soil. PFCs also have been used in firefighting foams.

It's no surprise that dogs absorb and carry PFCs and other chemicals from continual exposure to the goods we buy. In the EWG's study, blood and urine samples in particular contained carcinogens, chemicals toxic to the reproductive system, and neurotoxins. Dogs are thirty-five times more likely than humans to get skin cancer, four times more likely to get breast tumors, eight times for bone cancer, and twice as likely to develop leukemia than humans, so the presence of carcinogens is especially concerning. Later in the book, we'll share insights gleaned from more recent studies looking at the impact phthalates (a common ingredient in plastics) and lawn products have on our pets. These chemicals are ubiquitous in our environment and are playing into our health, whether we're aware of it or not.

People and their pets living in industrialized nations now have hundreds of synthetic chemicals in their bodies accumulated from food, water, air, and, let's not forget, contaminated dust and treated lawns. The vast majority of these chemicals have never been adequately tested for health effects. Most people don't realize that many everyday items, from food packaging, furniture, dog beds, and household items to clothing, cosmetics, and personal-care products, can contain harmful substances. These include the usual suspects, like pesticides, herbicides, and flame retardants, as well as phthalates, which are used to soften plastic; parabens, which act as preservatives; polychlorinated biphenyls

(PCBs), still found in the environment due to widespread use in a lot of electrical and cooling equipment; and bisphenols, which are pervasive in a wide variety of plastics, including plastic food and water bowls and lots of dog toys. Some of the most egregious chemicals can mimic or block hormones, disrupting vital body systems—hence, they are called "endocrine-disrupting chemicals" (EDCs).

Early-life exposure to bisphenol A (BPA), an ingredient in many plastics that makes its way into everything from bottles to toys to the lining of food cans, is linked to asthma and neurodevelopmental problems such as hyperactivity, anxiety, depression, and aggression. In adults, BPA exposure is associated with obesity, type 2 diabetes, heart disease, decreased fertility, and prostate cancer. Although BPA is often replaced with bisphenol S (BPS) and bisphenol F (BPF), these are less studied and may very well have similar hormone-disrupting effects. Prenatal and early-life exposure to phthalates is linked with asthma, allergies, and cognitive and behavioral problems. It may also affect reproductive development in males, both human and canine. In both species, phthalates are associated with reduced fertility. In fact, researchers have identified PCBs and other environmental chemicals in the gonads of neutered dogs. Obviously breeding is no longer an option for these dogs, but the thought of measurable amounts of environmental chemicals found in many of our dogs' organ systems shouldn't sit well with any of us.

Ironically, some of the studies that inform human health have come from looking at our pets. They can be early-warning systems for toxic chemical exposure at home. Case in point: Stockholm University's professor Åke Bergman took a novel approach to measuring blood levels of various chemicals in small children by turning to their pets. Rather than asking parents if he could test their children's blood, he took samples from their house cats, which spend much of their time on the floor and in a very similar environment to that of infants and young children. As the children are crawling and playing, the pets are swiffering the same floors and air. Bergman and his team found that there was a close association between levels of persistent organic pollutants they measured in house dust and those in the blood of house cats.

In 2020, a team of researchers out of North Carolina State and Duke University, who call the reality out in their paper's title— "Domestic Dogs Are Sentinels to Support Human Health"— employed a novel technology to show the shared chemical load. Using high-tech but inexpensive silicone wristbands and collars designed to measure exposures to chemicals in the environment, the group found remarkable similarities between dogs' and their owners' chemical loads. For example, they found a type of PCB in 87 percent of human wristbands and 97 percent of dog tags. This is surprising, considering that the US government banned the use of PCBs way back in 1979. But apparently, it can linger for decades and have far-reaching, often silent effects. Such research builds on previous work in other animals, including horses and cats. In 2019, environmental toxicologist Kim Anderson of Oregon State University, who helped develop the wristband technology, found an association between flame retardants and a disease in cats known as feline hyperthyroidism, an endocrine disease that has skyrocketed in the last forty years. That may be because cats like to rest (with us) on upholstered furniture, which often contains flame retardants. Remember, the EWG has found dozens of environmental chemicals in higher levels in pet serum than in human blood. We're hoping for a similar study looking at the effects of the chemicals sprayed on dog beds.

Countering the effects of this chemical soup, however, is easier than you think. We won't ask you to run out and buy all new, certified organic furniture and upholstery. You can work with what you've got and implement commonsense strategies like throwing an old cotton sheet or light natural-fiber throw blanket over your pet's bed and making sure they eat and drink from chemical-free bowls.

Although the Well Living Lab is not studying dogs in particular, we likely can draw similar conclusions because our dogs are living under the same conditions. They are exposed to the same pollution, which goes beyond classic industrial chemical exposure. Pollution can also come from noise and emanating light from screens late at night—all of which are insults to healthy biology. This includes the composition and biology of their microbiomes. This, in turn, impacts everything about us—our metabolism, our immune function, and ultimately our health and happiness. The same rings true for our dogs. Although each one of us houses unique microbiomes, patterns emerge when you consider cohabitating creatures (ahem: people and their pets). Indeed, we may share features of our microbiomes with our dogs and vice versa. That may not sound like a pleasant reality, but it's one that may confer better health and wellness for both human and canine.

Such an incredible two-way street is characteristic of many aspects of the human-dog dyad. The aging process is where this unique dyad shares the same one-way road. Let's go there next.

LONGEVITY JUNKIE TAKEAWAYS

➤ Dogs and humans have coexisted and comingled for centuries, and while the exact timeline of our best friends' domestication is still up for debate, all dogs evolved from the gray wolf and have enjoyed a special bond with humans. We have come to rely on each other, and we have so much in common when it comes to health.

➤ We have aggressively bred dogs over the past century for certain traits, and this has led to genetically weak canines. Genetics alone, however, do not necessarily determine a dog's health destiny. As with human health, environmental factors like diet, exercise, and exposures weigh heavily in the equation.

➤ Even though the list of core ingredients found in many commercial pet foods would suggest otherwise, dogs don't need a lot of sugar or starches, and there's a big difference between processed, refined carbs and fiber-rich, low-glycemic carbs that also aid in the health of the microbiome—the microbes in the gut that play into metabolism, mood, and immunity.

➤ Dogs are increasingly isolated, confined to indoor and potentially toxic spaces, and deprived of the outdoors, where access to activity, fresh air, microbially rich dirt, and the wilds of nature ultimately boost their health.

3

The Science of Aging

Surprising Truths about Dogs and Their Risk Factors for Disease

A bone to the dog is not charity. Charity
is the bone shared with the dog, when
you are just as hungry as the dog.

—Jack London

How old is my dog, really?

We get this question a lot. And we know what people are asking. If they could factor in their dogs' genetic vulnerabilities and overall health status to date, how long could their beloveds potentially live? Are they ahead of or behind their chronological age? It's like with humans: We all know people who look (and act) a lot younger or older than the calendar says. Some people seem to defy their age, while others show signs—inside and out—of accelerated wear and tear.

This is a picture of Augie jumping into the pool at sixteen years of age, something she did every day. Her dad, Steve, said she often wanted to continue retrieving the ball from the pool at this age, even after the other dogs were tired. Many Goldens die around ten and begin the process of physical degeneration, muscle atrophy, and sarcopenia (losing strength and function) prior to that. Augie clearly didn't follow the average Golden Retriever's life path. She

lived to be twenty years and eleven months, dying in the spring of 2021. She is currently the nineteenth-oldest dog ever and the world's oldest known Golden.

The whole concept of aging is extraordinary and has been the subject of debate for centuries, not to mention fodder for the humanities and the soul of scientific debate. Age is a number, a process, a state of mind, a biology, a condition, a reality, an inevitability, an onus, and a privilege. It's so many things yet nothing we can touch or feel. There are many theories about aging—what it means, how it works, where it starts, how it unfolds . . . and ultimately ends. While one person talks about the strength and length of chromosomes (especially those telomere tips that are like shoelace caps holding life together) or the integrity of cellular renewal processes, another focuses on DNA stability and repair mechanisms to keep up with mutations to prevent, say, cancer. The stability of proteins—the complex molecules that control nearly everything in the body directly and indirectly via structures, hormones, and overall signaling—also earns attention in aging circles. When there's a loss of proteostasis, which simply refers to "quality control" of the body's proteins and their related cellular pathways, trouble looms. *Proteostasis* comes from a combination of the words "protein" (a molecule that a cell uses as a machine or scaffolding) and "stasis" (meaning to keep the same).

A body, dog or human, loves stasis—equilibrium, stability, sameness day after day to maintain control. Harvard Medical School professor David Sinclair's study of one particular family of proteins called "sirtuins" has been particularly revealing. Sirtuins help control cellular health and serve a starring role in keeping cells balanced (homeostasis) and able to handle stress. They are thought to be responsible, in large part, for the cardiometabolic benefits of lean diets and exercise and when activated can delay key aspects of aging. But their actions depend on the availability of other important biomolecules such as nicotinamide adenine dinucleotide (NAD), a form of vitamin B. Ensuring a body is fully equipped to benefit from sirtuins is key, or things can start to break down and take a bite out of that longevity equation.

Then there's talk about inflammation ("inflamm-aging"), immune and mitochondrial dysfunction, stem cell depletion, free radicals and oxidation (biological rusting), miscommunication between cells, functional decline in the central nervous system—the list goes on and on. The mitochondria, for instance, are important tiny structures inside cells that produce energy; stem cells are like baby cells that can grow into *any* type of cell—as such, they are essential to cellular renewal and tissue regrowth. Free radicals (sometimes called "reactive oxygen species") are those rogue molecules that have lost an electron. You've no doubt heard the advertising campaigns from the health and wellness industry promoting free radical–quenching antidotes. Free radicals are instigators of trouble in the body. Electrons usually come in pairs, however, stress, pollution, chemicals, toxic dietary triggers, ultraviolet sunlight, and some bodily activities can dislodge electrons from a molecule. Then, that "free" electron may attempt to snag electrons from other molecules. Thus begins the oxidation process, as more free radicals form, which then produces inflammation. Because oxidized tissues and cells don't function normally, the process can render you vulnerable to a slew of health challenges. This helps explain why those with high levels of oxidation (oxidative stress), which is often reflected by high levels of inflammation, have an extensive list of health challenges.

You don't need to master these abstractions. You just need a general sense of why and how we age in order to make good decisions in your everyday life to optimize health and vitality. And how we humans age is how our dogs age.

Life is an ongoing cycle of destruction and construction, no matter what kind of creature you are. From the simplest chemistry in which molecules are pulled apart and rearranged to form new compounds, these life processes encompass the formation, growth, maintenance, and replication of the cell. They control both single cells and multicellular organisms, from yeast to dogs to the human body. Disturb too much of the functioning of either part of this process—the destruction or the construction—and the life process will become dysfunctional and, if not corrected, eventually end.

As you can imagine, age is the strongest risk factor for disease (the strongest predictor of health span); the older we are, the higher

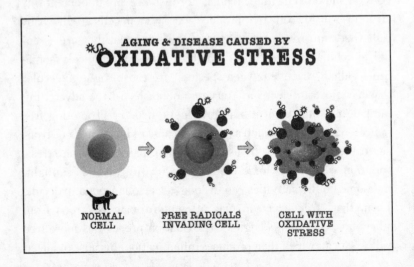

AGING & DISEASE CAUSED BY
OXIDATIVE STRESS

NORMAL
CELL

FREE RADICALS
INVADING CELL

CELL WITH
OXIDATIVE
STRESS

the risk for illness and the development of a degenerative disease. The same is true for our dogs, who age about six to seven times faster than we do (hence the customary multiplier of seven in calculating "dog years," though we'll see the imprecision of this math later). Because dogs live and die much faster than humans, it's easy to overlook how quickly the aging process occurs.

While different from humans in the sense that dogs live six to twelve times shorter than their people, their life circumstances—like a human's—change over time. Indeed, they have distinct development stages just like people, including puppyhood (from birth to between six and eighteen months), adolescence (between six and eighteen months), adulthood (starts between one and three years), the senior years (begins between six and ten years), and the geriatric phase (seven to eleven years). Another similar change with age is the nutrition requirements for dogs and humans influenced by activity level. It's really no surprise, not to belabor this point, that the rise in obesity among dogs (a rise of around 20 percent since 2007) mirrors the rise of obesity in humans. It's not always easy to picture a dog with cognitive abnormalities, but it happens more commonly than one would think: Almost one-third of eleven- to twelve-year-old dogs and 70 percent of fifteen- to sixteen-year-old dogs show cognitive disturbances with symptoms corresponding to human senile dementia: spatial disorientation, social behavior disorders (e.g., problems recognizing family members), repetitive (stereotypical) behavior, apathy, increased irritability, problems with sleep, incontinence, and reduced ability to accomplish tasks. Collectively, these symptoms are characteristic of the typical mental decline of aging dogs. It's called canine cognitive dysfunction syndrome, or "doggy dementia."

Because dogs and humans share similar developmental changes sparked by aging, canines provide a good representation for studying overall health. And get this: Dogs share more ancestral genomic sequence with humans than rodents do. In this endeavor, however, we learn that the causes of premature death in dogs are often the same for humans: a confluence of genetic and environmental pressures. This is one reason many scientists we interviewed are using dog models for human aging: They are sentinels, forecasters, and promoters of human health. We'll continue to explore these pressures and show how they combine to define a dog's life . . . and chances for a lusciously long and vibrant one.

Domesticated dogs have a lifespan that could range from five and a half to fourteen and a half years, depending on breed. There are several factors that determine how fast or slow a dog ages, including genetics, environment, and past experiences like trauma. There is no predetermined age when dogs will begin to decline because of age, and this decline will be influenced by weight, breed, size, and if there are any genetic predispositions to disease. The word "senescence" can refer either to cellular senescence or to senescence of the whole organism (the word can be traced back to the Latin *senex*, meaning "old," and gives us "senile" and "senior"). In recent years, the medical literature is rife with studies zeroing in on so-called zombie cells, which are senescent cells. These cells start out normal but then encounter a stressor, such as damage to their DNA or viral infection. At that point, the cells can choose to die or to become "zombies," basically entering a state of suspended animation in which they are not helpful to the body and hang around like moody, trouble-causing vagrants.

Issues begin when zombie cells let go of chemicals that affect normal cells they are close to. In mouse studies, drugs that delete zombie cells have been shown to improve an impressive list of conditions, such as cataracts, diabetes, osteoporosis, Alzheimer's disease, enlargement of the heart, kidney problems, clogged arteries, and age-related loss of muscle (sarcopenia). We also can target zombie cells by feeding specific nutrients, which we will cover in Chapter 8. This prospect helps explain why David Sinclair's work in turning

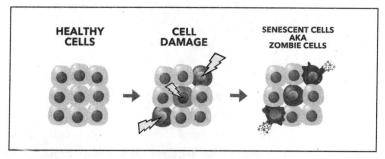

When healthy cells are diseased or damaged, they can stop dividing and become senescent "zombie" cells that spit out inflammatory molecules to trigger inflammation. As these zombie cells accumulate, nearby cells can also turn into zombies and speed up the entire aging process.

old mice into healthy, younger-acting mice has been so seductive to biotech pioneers.

Animal studies also have shown a more direct tie between zombie cells and aging. In one study, older mice were given drugs that targeted zombie cells, then placed on a treadmill. Results showed their walking speed, grip strength, and endurance improved. In addition, mice whose "human age" was between seventy-five to ninety years old were given these drugs, and their average lifespan increased by 36 percent. Conversely, research has shown that giving zombie cells to young mice causes them to act old, including slowing their walking speed and diminishing their strength and endurance. Furthermore, the implanted cells turned these mice's normal cells into zombie cells.

In simplest terms, cellular senescence refers to the phenomenon of old cells not dying. Like zombies in folklore, they are capable of movement but lack rational thought and do not respond appropriately to their surroundings. At some point, all cells stop dividing and should die so they don't clutter up the system and crowd out healthy new cells. If they stop dividing but don't die, they pave the way for problems in tissues, organs, and systems. This causes the biological system of checks and balances to move off track, resulting in an array of new risk factors for dysfunction and disease. The combination of lost stem cells and cellular senescence, for example, has been attributed to functional decline in the nervous system. The immune system can be affected considerably by miscommunication among cells, and this also could lead to increased inflammation in the body, called "inflamm-aging." Elevated levels of inflammation also may result from an increasing number of senescent cells.

Researchers have found that a natural plant compound (polyphenol) called "fisetin" reduces the level of these damaged zombie cells in the body. When older mice received fisetin treatments, their health and life span improved notably. Fisetin is present in many fruits and vegetables, including strawberries, apples, persimmons, and cucumbers. It also adds to the bright colors we see in our produce. You'll see in Part III that we encourage offering fisetin-rich fresh food treats and toppers to your dog's meal to add some of this antiaging biohack. When it's treat time, try serving up slivers of fisetin-rich strawberries and apples; in addition to enjoying a delicious snack, your dog will also get a dose of this powerful longevity molecule and gut-loving fiber, too.

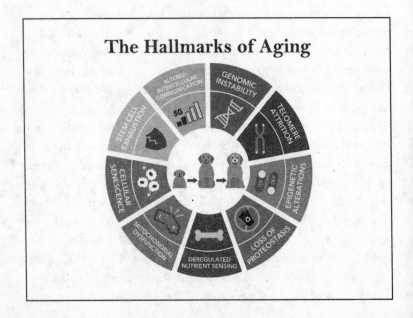

The Hallmarks of Aging

ALTERED INTERCELLULAR COMMUNICATION
GENOMIC INSTABILITY
TELOMERE ATTRITION
EPIGENETIC ALTERATIONS
LOSS OF PROTEOSTASIS
DEREGULATED NUTRIENT SENSING
MITOCHONDRIAL DYSFUNCTION
CELLULAR SENESCENCE
STEM CELL EXHAUSTION

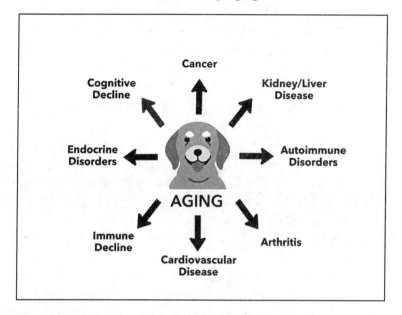

So you can see that the aging process is extremely complex and the contributing elements are exceedingly interrelated (which is represented in the diagram that follows). There is no single pathway. Multiple inputs contribute to the aging process and its speed—fast or slow.

You don't need to be a scientist to notice the signs of aging in your dogs—we generally spend the second half of our dogs' lives diagnosing and treating precisely these signs. Although one could argue that we all start aging at the moment of birth, it's widely thought that there's a pinnacle, if you will, around the age of twenty-five in humans (and around the age of three in a dog) that's particularly notable in the aging process. This is when certain biological events occur and put a body into the inevitable (and initially unnoticeable) descending slope of life, and those hallmarks of aging we just outlined become more possible and pronounced. In humans, cellular processes change, growth hormones shift, metabolism ticks down a notch, the brain reaches structural maturity, and muscles and bone mass peak. You may not feel or physically notice any of this until your forties or, if you're lucky, your fifties, but it begins in your twenties. Dogs have their own version of natural decline, too, and how fast it happens depends on lots of variables. In addition, the canine aging process can be deceiving because they often display high energy levels and appear super healthy, regardless of what's really happening internally.

Studies show that body weight is more predictive of life span than height, breed, or breed group, and big dog breeds do age more rapidly than small ones. This runs counter to what happens in other corners of the mammal kingdom. Large animals like whales and elephants have evolved to live long lives because they have no natural predators. These mammals can even evade cancer—more on this later.

But in the canine world, bigger is not always better. Big dogs like the Great Dane may only live to age eight, while smaller dogs like the Shih Tzu have life expectancies up to age sixteen. Since most dog breeds have only been around for a few hundred years, evolutionary pressure can be ruled out for the reason why there's such a discrepancy. Instead, hormones like insulin-like growth factor-1 (IGF-1), which increases in large dogs, could be the culprit. Studies point towards this protein being the reason for shorter life spans in several species, though how exactly IGF-1 works to impact age isn't fully known.

Kimberly Greer, PhD, who is a dog-loving geneticist at the Center for Canine Behavior Studies in Texas and teaches at Prairie View A&M University, was among the first scientists to connect serum IGF-1, body size, and age in the domestic dog more than a decade ago. When we spoke with her, she underscored that "size does matter" in our canine friends. Large dogs tend to die young; to win the longevity race, IGF-1 needs to be under control.

Interestingly, when the IGF-1 pathway carries a mutation, the result is a *longer* life span (this is the rare case where a mutation confers an advantage); put simply, lower levels of IGF-1 equate with longer life. Scientists have long documented this phenomenon in mice, flies, worms, and even humans. But the trade-off in mammals is often dwarfism because the mutation affects how the body uses growth hormone. Most remarkably, certain communities of people throughout the world who carry the IGF-1 mutation have very short stature (under five feet) but are protected from cancer and diabetes. This condition is known as Laron syndrome, named after Zvi Laron, the Israeli pediatric endocrinologist who first documented the disorder in 1966. Worldwide, there are between three hundred and five hundred individuals who have this peculiar disorder and continue to be studied.

This may be a reason behind why several toy breeds—Chihuahua, Pekingese, Pomeranian, and Toy Poodle—are much less likely to die of cancer than other breeds. Numerous animal breeds (e.g., miniature and teacup dogs, cats, and pigs) whose dwarfish size is due to a single mutation in their IGF-1 gene also enjoy remarkably longer lives than their normal-size ancestors. This is one example of a gene mutation yielding beneficial downstream effects.

We also know that body weight influences the onset of when age-related diseases like cancer take root, and this is true for dogs and humans. Larger dogs also tend to grow faster, which could result in what scientists at the Dog Aging Project call "jerry-built bodies" that are more susceptible to complications and disease. Throw in neutering before puberty, and there's another set of hormonal variables that impact health span and life span.

Let us pause here for a moment to address the issue of sterilization: If all of your hormone-producing organs (the ovaries or testicles) were removed before you hit puberty, you'd naturally worry about the long-term effects on your health and your risk for disease. Many women, for example, who have their uteruses removed for health reasons choose to keep their ovaries (if they can), so they can continue to gain the benefits from the important hormones produced there. The same logic holds true for dogs. Typical canine desexing (spaying and neutering) removes organs that researchers now believe are quite important to overall health in dogs. Studies also indicate that the earlier a puppy is spayed or neutered, the greater the likelihood of health problems later in life, from abnormal bone growth and bone cancer to an increased incidence of adverse reaction to vaccines and behavioral challenges like fear and aggression. My (Dr. Becker's) recommendation to pet parents who have not desexed their dogs yet is to consider a hysterectomy or vasectomy instead; these procedures lead to the same end result (sterilization) but without the negative physiological side effects.

In addition to hormonal factors in the health equation, size also appears to matter. Large dogs tend to have more health issues than smaller dogs. For example, German Shepherds frequently are diagnosed with hip dysplasia, and many Siberian Huskies experience autoimmune disorders. But some of these problems, as we'll later see, could also be the result of inbreeding and other factors that affect epigenetics.

The Dog Aging Project is one of many initiatives seeking to understand how genes, lifestyle, and environment influence aging. The focus is to understand dog aging by collecting and analyzing big data. Led by a consortium of scientists at top research institutions around the world and centered at the University of Washington and Texas A&M University, this long-term biological project, which we've personally had the privilege to learn more about from behind the scenes, aims to use the information gathered to help pets—and people like you and me—increase health span. Yet another example of how our leashes are a two-way street.

Additionally, a small component of the project explores the use of pharmaceuticals to potentially increase the life span of dogs. The project, which engages the general public to register their dogs in the studies and is the largest-ever study on aging in dogs, has funding from the National Institutes of Health. Elinor Karlsson, PhD, is among the scientists participating in the research from her base in Massachusetts, where she is the director of the Vertebrate Genomics Group at the Broad Institute of MIT and Harvard. Her Darwin's

Dogs project similarly recruits citizen scientists to share their pets' information so her research lab can draw connections between a dog's DNA and behavior. As she explained during our visit to her Boston office, her team hopes to identify causes to diseases and disorders— from cancer to psychiatric and neurodegenerative illnesses—that are in dogs' genetics, and those clues can lead to breakthroughs in treating the same conditions in people (all she needs is saliva samples from people's dogs and answers to basic questions).

Over at the National Human Genome Research Institute in Maryland, the Dog Genome Project, led by Elaine Ostrander, PhD, and a collaboration of international scientists, is building a database to understand canine genetics and their health implications. So far, the group has played a role in finding genes for retinitis pigmentosa, epilepsy, kidney cancer, soft tissue sarcomas, and squamous cell cancers—contributing powerfully to the concept of One Health medicine and to both the veterinary and human medicine literature. How so? The canine disease genes are often the same or related to those that cause disease in humans. In the dog realm, the collaborative effort has identified genes that contribute to variation in skull shape, body size, leg length, fur length, color, and curl. In their pursuits, scientists have assembled a genomic vocabulary for mammalian development. It's revolutionary science that will continue to unfold.

Many of the researchers behind all of these projects share their data and have professional co-partnerships. It's an all-hands-on-deck approach. Now that gene sequencing is quick, efficient, and relatively inexpensive, the speed of scientific discovery has accelerated. For example, researchers have identified more than 360 genetic disorders that occur in both humans and dogs, with approximately 46 percent of those occurring in only one or a few breeds. In one of the more famously documented correlations between the two species, scientists have located a mutation in the dog genome that causes narcolepsy. This finding led researchers to study the mutation in its human equivalent—proof that canine genetics can benefit humans.

Not surprisingly, **a dog's biggest risk factors for death are age and breed**. Age and the genetic cards one is dealt are the twin powers that define a life span. But what's not common knowledge is how various forces collaborate in subtle but influential ways to change the chances of premature death. Those forces include one's circadian rhythm, the state of one's metabolism and microbiome, the condition of one's immune system, and environmental effects that act on the genome.

RACE: The Stickiness to Aging

How many times have your dog's food ingredients been cooked? Once? Twice? So many times that you can't count? The answer to this key question is one way to assess the healthfulness of your dog's food, and we'll show you how to do that in Part III. Here's why this is important.

Insulin-like growth factor-1 (IGF-1), which we just defined, is an important protein. This protein's activity has an intimate relationship with two key hormones in all of us: growth hormone and insulin. Insulin, as you likely already know, is one of the body's most influential hormones, whether you're a dog or a human. It's a main player in metabolism, helping move energy from food into cells for their use. Because cells cannot automatically snatch glucose passing by in the bloodstream, they need the help of insulin, which is produced by the pancreas and acts like a transporter.

Insulin transports glucose out of the bloodstream and into muscle, fat, and liver cells, where it's converted into energy to power the body. Healthy cells do not have an issue working with insulin, as their cellular receptors for it are abundant. However, the number of insulin receptors your body has decreases after chronic exposure to high levels of insulin. In turn, this increases the amount of glucose in the system. This dynamic typically occurs because of consuming a large amount of refined sugar and simple carbs, especially from processed food. This causes our cells to become desensitized or "resistant" to insulin, ultimately causing insulin resistance and eventually type 2 diabetes, or lifestyle-induced diabetes (you weren't born with a faulty pancreas).

A majority of dogs diagnosed with diabetes are not born with the condition, or if they are, it's caught when they are puppies. Diabetes comes along later after the pancreas stops producing the level of insulin needed and the cells become diminished beyond repair. The end result is too-high levels of blood sugar outside the cells, instead of inside the cells making energy. If too much sugar (glucose) remains in the bloodstream, that sugar will inflict a lot of damage, including creating advanced glycation end products (aptly named AGEs), in which "sticky" glucose molecules attach to proteins (like those that make up your inner blood vessels) and cause dysfunction. The major receptor for AGEs, known as the receptor for advanced glycation end products, is aptly called RAGE!

Glycation (as the AGE process is called) occurs any time there is heat, glucose, and protein found together; it's a chemical reaction that occurs both inside and outside our bodies. When it occurs inside our bodies, it causes premature aging and inflammation in dogs and humans. We'll go into more detail about all this later, because in addition to the AGEs made by the body, they're also found in foods that have been processed using heat. This is one of the factors we'll ask you to consider when choosing your dog's food, going forward. **When glycation occurs in food processing, it's referred to as the Maillard reaction, and the end result is Maillard reaction products (MRPs).** When we consume or feed food that contains MRPs, we compound our burden of having to deal with these toxic substances: we eat them *and* our bodies make them. Even worse, a second type of MRP occurs when dietary fats are heated with protein, resulting in lipid peroxidation, the production of toxic substances called "advanced lipoxidation end products" (ALEs), which also bind to the same receptors, inciting more RAGE.

The F-word (fat) is something we've come to fear: Fat is bad. But just like carbs, there are good fats and bad fats. No doubt, rancid, oxidized, heated fats wreak havoc on health, creating cytotoxic, cell-damaging compounds that have a wide range of downstream functional consequences in the body, from pancreatitis and liver dysfunction to immune dysregulation. Dogs need clean, unadulterated sources of fats and fatty acids to survive and thrive, and they must consume adequate amounts to avoid disease and maintain health. Dogs need fats for supporting healthy brain chemistry, fostering optimal skin and coat health, absorbing certain nutrients, and producing important hormones, among many other functions. Consuming the rendered, oxidized, high-heat-processed pet food fats means that dogs are consuming vast amounts of toxic ALEs.

A 2018 study out of the Netherlands found dogs *consume up to 122 times the AGEs* in their diets as humans. This rattled the proactive veterinarian (Dr. Becker) in me to the point of not sleeping. I contacted board-certified veterinary nutritionist Dr. Donna Raditic and asked if we could design and fund a study evaluating AGEs across the most popular pet food categories: raw, canned, and dry dog food. Dr. Raditic and I co-founded the Companion Animal Nutrition and Wellness Institute, a nonprofit organization, for the purpose of conducting unbiased, university-based companion animal nutrition research, because basically there isn't any. The Big Five pet food companies conduct some internal research that never reaches the general public, but there's no government-funded National Institutes of Health for pets. No one is funding very important basic nutrition research that allows veterinarians to know more about how food impacts health and disease, or what the consequences or benefits are to the arbitrary feeding trends not rooted in any science, like feeding ultra-processed "fast food" to pets for decades and telling people it's healthy and the best possible choice.

We don't know if it's healthy because there's not one study comparing a group of dogs that eat real food their whole lives to a group that eats ultra-processed food their whole lives. We don't have any basic research that answers the commonsense questions many pet parents have, and in one regard we think the industry wants to keep it that way. What happens if we measure the levels of AGEs, mycotoxins, glyphosate, or heavy metals in pet food and discover they far exceed the level thought safe for humans? Some advocacy groups have done this on a small scale, testing a handful of popular brands for contaminants, with scary results that have sometimes resulted in pet food recalls.

Food Fights

Take a look at the graphic below, which shows the pet food recalls (in pounds of food) from 2012 to 2019:

Kibble and treats made up about 80 percent of the total pounds of pet food products recalled. The top four causes of recalls based on pounds of food were bacterial contamination (Salmonella), toxic levels of synthetic vitamins, unapproved antibiotics, and pentobarbital (animal euthanasia solution) contamination. The "other" category includes freeze-dried, cooked, and frozen diets and toppers. But recalls include only products that have been tested for a handful of accepted problems, most commonly excessive levels of synthetic vitamins or minerals and potentially pathogenic bacteria. Because the FDA doesn't require companies to test for other types of foodborne toxins, pet foods aren't recalled for shocking levels of glyphosate (Roundup) or AGEs.

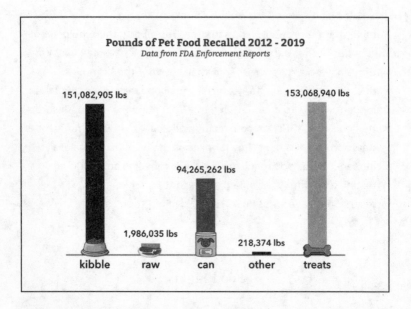

Pounds of Pet Food Recalled 2012 - 2019
Data from FDA Enforcement Reports

151,082,905 lbs
153,068,940 lbs
94,265,262 lbs
1,986,035 lbs
218,374 lbs

kibble   raw   can   other   treats

No spoiler alert needed: Less-processed foods are healthier for your dogs.

When the levels of AGEs in different dog food categories were compared, AGEs were highest in canned food, followed by dry food. Not surprisingly, the lowest amount of AGEs were found in minimally processed raw foods. The research behind the negative health consequences of consuming high levels of MRPs (AGEs) is undeniable. Because of this, it's important to investigate how many and how often the ingredients in the pet food you choose have been heat-adulterated.

In the human food space, the best example of all-in-one food is infant formula milk replacers. In theory, both pet food and infant formula are "nutritionally complete" diets. In the 1970s, Nestlé (also the owner of Purina) convinced millions of women to replace their breast milk with Nestlé infant milk replacer, declaring that it was healthier than breast milk and better for babies. Many women followed this recommendation, dumping their own breast milk and using an all-in-one powder instead. This marketing campaign prompted health advocates around the world to melt down, which led to massive public outcry and boycotts, several lawsuits, and a global education campaign about the health benefits of human breast milk. The same revolution is now occurring in the pet food industry: Animal advocates are demanding real, whole-food meals, not pelletized sustenance from food powders blended with synthetic vitamins and minerals.

The concept of all-in-one pet food wasn't always popular. In fact, pet food is a relatively recent addition to the marketplace, courtesy of one highly driven entrepreneur.

Dog Cakes and Milk Bones

One upon a time, there were no dog biscuits or milk bones. Someone had to invent them, and in 1860 James Spratt was the first to manufacture a dried dog biscuit food. His qualifications—he was an electrician and lightning rod salesman from Ohio—were dubious at best. He did, however, have the slick sales skills to turn a chance observation idea into a superbly contrived fortune, appealing largely to the elite at the start. While on a business trip to England, Spratt observed street dogs eating ship hardtack, the nonperishable, cracker-like food made from processed cereals that sailors ate on long trips (and which were often filled with worms, earning the nickname "wormcastles" among soldiers who ate it during the Civil War).

The idea for commercial dog food was born. Spratt called his original biscuit the "Patented Meat Fibrine Dog Cake"; it was made from a mixture of ingredients such as wheat, beetroot, and various other vegetables that were bound together using beef blood and then finally baked. We may never know what was in the original biscuit, which included "the dried unsalted gelatinous parts of Prairie Beef." While he was alive, Spratt remained very secretive about where the meat in his biscuits came from.

At a hefty price point, the price equating to a day's wages for a skilled craftsman, Spratt targeted an audience of wealthy English gentlemen. His company began operations in America in the 1870s, and Spratt targeted health-conscious pet owners and dog-show participants, buying the full front cover of the first American Kennel Club journal in January 1889. Americans quickly subscribed to the rhetoric and started buying biscuits instead of giving their pets leftovers. In addition, Spratt launched the idea that dogs at different ages should be eating different foods. Spratt wins points for his marketing smarts: his company was the first to erect a billboard in London, and he enticed his wealthy friends to review and and promote his dog cakes.

After Spratt's death in 1880, the company went public and became known as Spratt's Patent, Limited, and Spratt's Patent (America) Limited. The establishment was not about to die. Much to the contrary, Spratt's became one of the most heavily marketed brands in the early twentieth century, with product recognition developed through logo display and lifestyle advertising, and support through devices such as cigarette cards. In the 1950s, General Mills acquired Spratt's US business. James Spratt's story is one of classic American entrepreneurism. He may come across as a hero of sorts—a purveyor of fine dog treats for health-conscious pet owners. But don't be fooled: Fundamentally, Spratt was a savvy, financially motivated salesman at the right place at the right time. He saw and leveraged opportunity in an environment entirely devoid of easy, convenient pet food solutions. His idea eventually grew into a multibillion-dollar pet food industry. Some of the marketing claims drummed up more than a century ago are still used (effectively) today. Indeed, it wasn't long before competitors used Spratt's own playbook to provide customers an even wider assortment of pet food convenience, marketed and purchased as a sign of our love for and commitment to our pets.

In 1948, veterinarian Dr. Mark Morris partnered with Hill Rendering Works to create the first "prescription" pet food diets. To this day, the term "prescription diet" is misleading: there aren't any medications or special substances included in these diets; they're called "prescription" because they're sold only by veterinarians. These popular diets weren't always so popular. In fact, Hill's Science Diet struggled to be profitable in the 1970s, until toothpaste giant Colgate-Palmolive purchased the company and wooed customers by using pages from their own expert brand-ambassador marketing strategy. Colgate-Palmolive had struck gold when it had dentists smile while holding tubes of toothpaste, declaring "the brand most dentists recommend." Colgate toothpaste became an instant top seller after the company's marketing team decided to try this novel marketing approach. Now the company knew how to strike gold . . . if it worked for dentists and toothpaste, why not vets and dog food?

Soon enough, Science Diet replicated its arrangements with dental schools in the veterinary world, striking contracts with vet schools and even funding nutrition professorships. Today, every veterinary school has partnerships with one of the top five pet food brands. A few good questions: What happens when medical and vet schools have exclusive alliances with industries like pharmaceutical companies and food manufacturers? Is this not a patent conflict of interest? Do these alliances seed bias in the schools' research and the education of their students?

Pet nutrition research differs from human nutrition research in lots of ways. For pets, the first published data for minimum nutrient requirements came in a 1974 book from the National Research Council (NRC), *Nutrient Requirements for Dogs and Cats.* It was a culmination of all of the pet food research completed using small groups of research puppies and kittens living in laboratories eating mid-century kibble with study designs that would no longer pass any university ethics committee. This book remains the definitive guide that the Association of American Feed Control Officials (AAFCO) uses to set nutrition parameters for pet food manufacturers. The go-to NRC reference has been updated once, in 2006.

All sorts of corporate marriages and divorces have occurred since the budding pet food industry was born a little over one hundred years ago. Nestlé bought Purina in 2001 and makes more than twenty consumer brands, such as Alpo, Beneful, Dog Chow, and Castor & Pollux. The company that dominates the market in the number one slot remains Mars Petcare Inc. (yes, the same company that makes a lot of your Halloween candy), which owns Royal Canin therapeutic diets and twenty-eight mass-market pet food brands, including Pedigree, Iams, and Eukanuba. Hill's Pet Nutrition currently stands at number three and is nearly tied with J. M. Smucker, which brings us Milk-Bone, Snausages, and Pup-Peroni. By the time this book prints, there will be more changes, as pet food is big business and a hot commodity for profit-focused multi-national corporations.

Interestingly, many nutrition advocates have launched public outcry and awareness campaigns against all-in-one pet food over the years, similar to the breast-milk awareness campaigns. There have been several fresh pet food pioneers over the years—Juliette De Baïracli Levy, Dr. Ian Billinghurst, and Steve Brown, to name a few—who have steadfastly maintained that pellets can never replace a dog's evolutionary diet. This was the concept behind the rise of the fresh pet food movement, which has become one of the fastest-growing segments of the pet food industry. Fresh pet food advocates have myriad other "issues" with the

TOP 10 U.S. PET FOOD COMPANIES ANNUAL REVENUE IN 2019

1. **Mars Petcare Inc.** - US$18,085,000,000
   Brands: Pedigree, Iams, Whiskas, Royal Canin. Banfield pet hospitals, Cesar, Eukanuba, Sheba, and Temptations.

2. **Nestlé Purina PetCare** - US$13,955,000,000
   Brands: Alpo, Bakers, Beggin', Beneful, Beyond, Busy, Cat Chow, Chef Michael's Canine Creations, Deli-Cat, Dog Chow, Fancy Feast, Felix, Friskie's, Frosty Paws, Gourmet, Just Right, Kit & Kaboodle, Mighty Dog, Moist & Meaty, Muse, Purina, Purina ONE, Purina Pro Plan, Pro Plan Veterinary Diets, Second Nataure, T-Bonz and Waggin' Train, Zuke's, Castor & Pollux

3. **J.M. Smucker** - US$2,822,000,000
   Brands: Meow Mix, Kibbles 'n Bits, Milk-Bone, 9Lives, Natural Balance, Pup-Peroni, Gravy Train, Nature's Recipe, Canine Carry Outs, Milo's Kitchen, Snausages, Rachel Ray's Nutrish, Dad's

4. **Hill's Pet Nutrition** - US$2,388,000,000
   Brands: Science Diet, Prescription Diet, Bioactive Recipe, Healthy Advantage

5. **Diamond Pet Foods** - US$1,500,000,000
   Brands: Diamond, Diamond Naturals, Diamond Naturals Grain-Free, Diamond Care, Nutra-Gold, Nutra-Gold Grain-Free, Nutra Nuggets Global, Nutra Nuggets US, Premium Edge, Professional and Taste of the Wild. Also: Bright Bites snacks.

6. **General Mills** - US$1,430,000,000
   Brands: Basics, Wilderness, Freedom, Life Protection Formula, Natural Veterinary Diet

7. **Spectrum Brands/United Pet Group** - US$870,200,000
   Brands: Iams (Europe), Eukanuba (Europe), Tetra, Dingo, Wild Harvest, One Earth, Ecotrition, Healthy Hide

8. **Simmons Pet Food** - US$700,000,000
   Brands: 3,500 SKUs, mostly in the wet pet food space.

9. **WellPet** - US$700,000,000
   Brands: Sojos, Wellness Natural Pet Food, Holistic Select, Old Mother Hubbard Natural Dog Snacks, Eagle Pack Natural Pet Food, Whimzees Dental Chews

10. **Merrick Pet Care** - US$485,000,000
    Brands: Merrick Grain Free, Merrick Backcountry, Merrick Classic, Merrick Fresh Kisses All-Natural Dental Treats, Merrick Limited Ingredient Diet, Merrick Purrfect Bistro; Castor & Pollux ORGANIX, Castor & Pollux PRISTINE, Castor & Pollux Good Buddy; Whole Earth Farms; Zuke's

These 10 pet food companies ranked among the top in 2019 annual revenue among companies based in the United States, according to Petfood Industry's Top Pet Food Companies Current Data.

ultra-processed, all-in-one diets made by the pet food industry,
including the following:

> Pet food packaging does not include a nutritional panel,
  similar to the labels on human food, that identifies the
  quantity of nutrients in the food, including how much sugar
  (or starch) is in the food.
> AAFCO owns the pet food ingredient definitions listed
  on pet food products. To find out AAFCO's definition of
  "chicken," you must buy its official publication for $250.00
  (spoiler alert: pet food chicken is *not* the same as grocery
  store chicken).
> Digestibility studies are optional.
> No batch testing is required for nutritional adequacy,
  contaminants, or toxins.
> In the United States, AAFCO sets the minimum requirements
  for pet foods to be considered nutritionally complete, but only
  a few nutrients have maximum thresholds. This means it's
  acceptable to produce a pet food with excessive amounts of
  other nutrients that can cause organ damage.
> Multiple studies validate that many pet food products are not
  accurately labeled, with higher-cost ingredients or proteins
  replaced with cheaper, unlabeled ingredients.

It's no secret why ultra-processed foods are so popular. They are
convenient, just like fast food is for human consumption. But we
have to ask: How much have we sacrificed health for convenience?
Just as there's been a push for more wholesome, minimally processed foods in the human space, there's now a big push for leveraging the power of real food in the dog space—especially given the
science of how strongly nourishment plays into the aging process.
Better nutrition means less stress on the body, and less stress on
the body means a longer life.

The Three Bones of Aging and Decline

Every creature on this planet is under constant pressures that culminate in aging. How any one of us ages is the result of three powerful forces in particular, no matter which way you look at it: (1) direct genetic effects from the DNA you inherited from your biological parents; (2) indirect genetic effects from how your DNA actually behaves, as we outlined at the top of the chapter; and (3) direct and indirect effects from the environment (diet, exercise, chemical exposure, sleep, and so forth). These influences are complex, interactive, and dynamic. Put together, they explain the "whys" of your individual health status and whether or not you live to one hundred. Your DNA, your DNA's behavior, and your exposures in the environment are unique to you and only you. The same goes for your dog.

Earlier in the chapter, we explained that your DNA, and how it behaves, is constantly at the mercy of environmental forces. A simple way to understand this is to consider a rescued dog you bring into your home. He's underweight, frail, out of shape, and apprehensive from being abandoned on the street. You nurse the dog back to health, and within a few months he's the very portrait of vitality—playful, fit, and confident. That dog still has the same underlying DNA, but clearly the genes are expressing themselves very differently due to a dramatic change in the environment. He's found a home in which he's well fed and loved. And it works both ways: An animal (be it a dog or a human) can have no genetic risk factors for a health condition and still develop one by virtue of daily habits. We all know people who are diagnosed with diabetes or cancer, for example, who have no history of these ailments in the family. Dogs similarly can develop a health challenge with no genetic history of it.

This is when *epigenetic* forces come into play.

One of the most intriguing areas of research today is in epigenetics, the study of particular sections of your DNA, or genome, that provide your genes the information about how strongly and how often to express themselves. It may help to think of these highly important sections as traffic signals for your genome. They flash stop and go instructions to your DNA and essentially act as a controlling mechanism for your health, lifespan, and what genes will be handed down to your descendants. Our day-to-day lifestyle choices have a profound effect on the activity of our genes. We now know that the food choices we make, the stress we experience or avoid, the exercise we get or neglect, the quality of our sleep, and even the relationships we choose actually choreograph to a significant degree which of our genes are "on" or "off." And the best news is that we have the ability to adjust the expression of several genes that impact our longterm health. The same is true for most of our canine companions, with one caveat: *It's up to us to make wise decisions for them.* If you're like us, this feels like a tremendous amount of pressure, especially when human and animal medical doctors currently aren't trained to be proactive health coaches capable of designing bio-individualized wellness protocols for us or our family members. Until there's a medical paradigm shift, the responsibility for knowing enough to make great choices for our own bodies, and our animals, is up to us.

Cellular Danger Response:
How Cell Trauma Speeds Up Aging in Young Dogs

You can't get through life without some damage, such as exposure to environmental chemicals, infectious diseases, or physical trauma. As David Sinclair often says, "Damage accelerates aging," but how? When damage happens, the damaged cells go through three stages of healing called the "cell danger response" (CDR). The process becomes less efficient as dogs age, and incomplete healing results in cell senescence and accelerated aging. For this reason, science now points toward interruptions of healing at the molecular level, early on in life, as viable contributing root causes of accelerated aging and chronic disease.

The three stages of recovery, which are controlled by the cell's mitochondria, must be completed successfully after the cell has encountered stress or a chemical or physical insult, otherwise, dysfunctional cells ultimately lead to dysfunctional organ systems. In other words, if cell healing is incomplete between injuries, more severe disease is produced. Chronic disease results when cells are caught in a repeating loop of incomplete recovery followed by re-injury, and are unable to fully heal. We don't have the statistics on dogs yet, but as dog owners, you'll recognize the symptoms of incomplete cellular recovery, triggering systemic disease: chronic allergies, organ diseases, musculoskeletal degeneration, immune system imbalances (from chronic infections to cancer). This is one microscopic, common, and quiet way intracellular disease begins when your dog is young and appears healthy on the outside.

Cellular Switches

Another area of focus pertaining to our ability to control the rate of aging and cell division is the evaluation of internal signaling pathways. One genetic nutrient-sensing "switch" that has been a center of attention in research circles lately is mTOR, short for mechanistic (formerly known as mammalian) target of rapamycin. You can think of mTOR simply as the school principal in all of our cells (except blood cells). When we interviewed Dr. Enikő Kubinyi, who is principal investigator of the Senior Family Dog Project at Eötvös Loránd University in Budapest, she was quick to point out that comparing the genetic pathways of aging between humans and dogs is similar and involves some of the same biomolecules, such as mTOR and AMPK (short for adenosine monophosphate-activated protein kinase).

AMPK is an antiaging enzyme. When activated, it promotes and helps control an important pathway called "autophagy" (self-eating) that basically manages cellular housecleaning and, in turn, enables cells to act in a more youthful manner. This pathway serves a lot of diverse roles in the body, but fundamentally it's how the body removes or recycles dangerous, damaged parts, including those useless zombie cells and pathogens. In the process, the immune system gets a boost and there's a greatly reduced risk of developing cancer, heart disease, autoimmune disease, and neurological disorders. The molecule is key to the cells' energy balance, too. We also know that AMPK may activate our innate "antioxidant genes," which are responsible for the body's natural production of antioxidants. As we'll see later on, it's much better to activate the body's internal antioxidant system than to take antioxidant supplements.

As is the case for humans, this metabolic pathway in dogs signals growth and cellular differentiation (i.e., whether a cell becomes a member of a muscle or the eye) and can be turned up or down like a dimmer switch by lifestyle factors such as diet, meal timing, and exercise. When you fast, for example, mTOR is suppressed and AMPK cleans house, which partly explains the benefits to fasting or, in the case of dogs, time-restricted eating (TRE; see Chapter 4). It's also about controlling blood sugar, because reducing insulin and those IGF-1 levels is related to turning down mTOR and turning up autophagy. When you live a sedentary lifestyle coupled with a diet consisting mainly of processed and inflammatory foods, it packs a punch to your cells not to mention skews insulin flow and blood sugar levels. This is a very simplified way to explain autophagy, but it's one of the key players in the aging process (and life itself). It helps to know that we all have this built-in technology to renew our cells and strengthen their performance. By following the strategies in this book, you'll help activate this biological Swiss Army knife in your dog.

RAPAMYCIN: FUTURE MEDICINE?

The "R" in mTOR, as noted above, represents rapamycin, a compound that's actually produced by a bacterium. It was named in honor of the site where it was first discovered in the early 1970s: Easter Island, otherwise known as Rapa Nui, which lies more than two thousand miles off the coast of South America. (Today it's a World Heritage Site owned by Chile, famous for its archaeological sites, including nearly nine hundred monumental statues called "moai" created by inhabitants during the thirteenth to sixteenth centuries.)

Rapamycin acts similarly to an antibiotic, with powerful antibacterial, antifungal, and immunosuppressive effects. In the early 1980s, labs began studying rapamycin, and over the next decade a stream of scientific papers came out reporting its effects on cell growth in yeast, fruit flies, roundworms, fungi, plants, and, most important for us, mammals. It wasn't until 1994 that scientists finally discovered the mammalian version of TOR, thanks to the work of Dr. David Sabatini and his colleagues at the Johns Hopkins University School of Medicine in Baltimore and the Memorial Sloan Kettering Cancer Center in New York. It helps to think of mTOR functions as the central hub of the cell signaling system, the command and control center of the cell. There's a reason why it's been conserved through 2 billion years of evolution: a master regulator of cell growth and metabolism, it's one of the secrets of how cell metabolism—life—is orchestrated within the cell.

Today, FDA-approved rapamycin is used in organ transplant patients to prevent rejection, and it has become one of the hottest antiaging and anticancer drugs under investigation. Ambitious studies on the use of rapamycin in dogs is already underway, and it will be exciting to see what's discovered that will inform and improve not only dog health but human health as well. To be clear: we are not "prescribing" rapamycin for you or your dog, but this latest science is worth mentioning because you'll no doubt be reading about it in mainstream media for years to come. The good news is you can influence mTOR with diet and lifestyle.

A Note about Cancer

Fear of a cancer diagnosis weighs heavily on many people's minds. Dogs manifest a broad range of cancers such as melanoma, lymphoma, osteosarcoma, and soft tissue sarcoma, as well as prostate, mammary, lung, and colorectal carcinomas. Approximately one in three dogs will be diagnosed with cancer during its lifetime; half of the dogs over ten years of age die from or with it. Much of what is known about canine cancers closely parallels what is known about human cancers.

Cancer is a complex disease in both humans and dogs. It's partly genetic, but not all mutations that result in cancer are heritable, and there are plenty of theories in the story of cancer, including the impact of mitochondrial damage to the process. For now, let's focus on the genetic aspect. A dog's DNA may change over the course of their lives. Genetic mutations can occur, and if one cell accrues enough mutations of a crucial gene, then those cells could begin to divide or grow in earnest. The cell will then cease to perform its designated function, and this may lead to cancer. This is called the Somatic Mutation theory, and in the last ten years, this hypothesis has been questioned by a growing number of oncology researchers who argue that cancer is a mitochondrial metabolic disease. Recent cancer research demonstrates that if a cancerous nucleus is transplanted into a normal cell, the cell remains normal. However, if you transplant the mitochondria from a cancerous cell into a normal cell, the cell becomes cancerous.

The metabolic theory of cancer, then, entertains the idea that there are things we can do to influence the health and well-being of our mitochondria, which can influence our cancer risk and, if needed, cancer treatment. Regardless of what cancer theory you subscribe to, the end result is the same—mutated DNA. Some powerful genes have been identified that can start the process themselves, often with a simple mutation. The breast cancer susceptibility genes BRCA1 and BRCA2 fall into this category. Missing genes also can cause an increase in cancer susceptibility. Bernese Mountain Dogs and Flat-Coated Retrievers are breeds that more commonly experience important tumor suppressor gene deletions CDKN2A/B, RB1, and PTEN that predispose them to histiocytic sarcomas. Finally, environmental factors contribute to cancer as well, such as smoking and lung cancer in humans, and lawn chemicals and lymphoma in dogs.

Whether the root cause is genetic, environmental, or a combination of both, a cancer diagnosis usually occurs when uncontrolled growth forms a huge number of abnormal cells that doctors call neoplasia. Neoplasia usually results in masses or tumors and is the product of a long string of dysfunctional occurrences in the body, beginning with the failed cell danger response to healing and ending with a completely confused cell containing dysfunctional mitochondria and permanently damaged DNA. Each cell contains a copy of the mutated gene identical to that from the original mutated cell. Tumor cells can migrate to other organs and begin to grow there. This is called "metastasis." The goal of cancer therapy is to kill all tumor cells within an affected individual, since a single remaining cell may cause the cancer to recur. Radiation treatment is used as a "local therapy" and is aimed at killing cells within the tumor site itself. In a similar manner, surgery often is used to remove the tumor. Chemotherapy is a "systemic therapy" that kills rapidly growing cells, in both the tumor and, hopefully, those cells that have traveled to other organs (metastasized). Chemotherapeutic agents also kill rapidly growing *healthy* cells; that's the problem.

To date, treatment for most diseases (including cancer) are undertaken retrospectively, once the disease is diagnosed. Since dogs are unable to communicate about their health, this typically happens at an advanced stage. Hopefully new genetic findings will change this. Thankfully, simple diagnostics such as the Nu.Q Vet Cancer Screening Test are now available in North America. Indeed, one of the most exciting possibilities in studying cancer lies in the ability to use genomics to identify mutations and diagnose cancer before it becomes a major problem. Ultimately, we hope to produce genetic tests to identify deleterious mutations before a dog gets sick. And the scientific community hopes to work with the canine breeding community to deplete disease susceptibility gene profiles from the populations. Dr. Brenda Bonnett of the International Partnership for Dogs is striving to do just that.

Happy, Healthy Dogs

Enhancing health span boils down to avoiding (or at least delaying) three categories of decline: cognitive, physical, and emotional/mental. We can't overstate the third bucket because it's often downplayed or ignored. We all know that unrelenting stress is toxic to our health, but what about a dog's stress levels? No joke: A Finnish study shows 72.5 percent of dogs display at least one form of anxiety, among other psychiatric issues we often attribute to humans, such as compulsive behaviors, fears and phobias, and aggression. Don't for a second think this is not a big deal. Studies of dogs who suffer psychological injuries or were poorly socialized early in life, for instance, point to serious long-term effects in their health and longevity. Once again, such studies mimic those done on humans. Trauma and fear are crippling to our long-term well-being. This is a silent but deadly problem. Millions of these dogs end up in shelters every year, euthanized for "behavior problems" that were prompted by unmanaged experiences and events early in a dog's life and exacerbated by inappropriate and oftentimes abusive "corrective" training measures. Ultimately, unaddressed emotional trauma robs us and our companions of a vigorous, joyous life.

Many pets are given psychiatric medications; according to a 2017 national survey by a market research firm, 8 percent of dog owners and 6 percent of cat owners gave medications to their pets for anxiety, calming, or mood purposes within the previous year. That translates to millions of animals in the United States taking medications for behavioral issues. A 2019 UK survey of dog owners revealed 76 percent of people wanted to change one or more of their dogs' behaviors. The Finnish study we just highlighted found that 32 percent of the 23,700 dogs tested experienced noise sensitivity, suggesting that it is the most common manifestation of anxiety. Some versions of human medications have received approval by the Food and Drug Administration for specific mental-health uses in pets, including the antidepressant clomipramine (Clomicalm) for separation anxiety in dogs and the sedative dexmedetomidine (Sileo) for dogs with noise-aversion problems. Most frustratingly, behavior-altering drugs don't produce a dog with a brand-new calm and balanced disposition. Behavior-modifying drugs don't give your dog a new personality; you're still left coordinating and executing behavioral interventions that help manage your dog's stress response.

Sadly, the truth is that many of these dogs lack proper mental and environmental stimulation; fear-free, relationship-centered training; and social connection. This is an ethical quagmire, particularly given the fact that we're talking about patients who cannot speak their minds in our language, whose behaviors are often misinterpreted, and who are required to learn a foreign language to know what we expect of them.

Every dog trainer and behaviorist reading these words will agree that many of the behavior problems we see in dogs today are a direct result of poorly or inadequately socialized puppies growing up in

homes with inadequate daily exercise, ineffective two-way communication (learning to understand your dog), and insufficient dog-chosen
hobbies or interests ("dog jobs"). According to renowned puppy trainer
and behaviorist Suzanne Clothier, who has been working with animals since 1977, "The formative experiences that impact us the most
throughout our lives are those that occurred when we were young
and impressionable: it's the same for dogs." Were our early childhood
experiences safe, predictable, and enjoyable? Or were they scary, unpredictable, lonely, or painful? The answer matters very much.

I (Dr. Becker) have had so many clients tell me they grew up
in dysfunctional homes and would never want to intentionally repeat the parenting mistakes that occurred during their childhood,
yet they find themselves parenting their dogs the way they were
parented—especially when it comes to physical roughness, patience during conflict, behavior challenges, and yelling when they
get frustrated. Dogs are as helpless, dependent, and powerless as
we were as children: vulnerable, not capable of changing their circumstances, faced with confounding language barriers, and unable
to effectively communicate emotions. We may have been too young
to verbalize how we felt or unable to use words to express what
we were thinking, but we felt intense emotions that may have been
overlooked or went unacknowledged, nonetheless. Fear. Anxiety.
Frustration. Confusion. Your dog has those same emotions, plus
massive language and social differences.

It's completely normal for dogs to growl when they feel scared or
threatened; they are communicating. Yet most people tend to punish
puppies when they growl. Training and teaching are two different
approaches to help your dog learn what you're asking for, what you
want. Training doesn't consider the animal's individual emotional
experience, whereas teaching takes into account the best way for the
student to learn. Just as children learn and process information in
different ways, so do dogs. It's our responsibility, as teachers, to instruct our kids in a way they can comprehend and respond to. Imagine being adopted by foreigners when you were young, then being

screamed at in a strange language or physically punished when one of them tried taking things from you and you merely defended yourself. Welcome to growing up in the confusing world of domesticated dogs: We expect them to act like human children who went to finishing school to learn perfect manners and got a PhD in good behavior when we haven't done our part to get them through even one Montessori class of puppy kindergarten, much less pull our weight in the homeschooling or active listening departments.

You can't rely on an hour of preschool each week to teach kids all they need to know about the world. As parents, if we aren't intentionally teaching at home on a day-to-day basis, our kids will not learn how to listen to us or communicate with us. Likewise, without a commitment to ongoing dog homeschooling, our adorable bundles of cuteness will make up their own rules according to their own culture, settling on behaviors that will not be socially acceptable to us in a year or two down the road. Building a solid relationship with your dog takes time, trust, consistency, and excellent two-way communication. But the payoff is huge: a better-behaved dog that has less stress and, in turn, a longer life.

We'll cover more about getting off on the right training foot later. For now, let's bust another myth that has everything to do with aging and, in particular, "age" itself: that every seven human years is equivalent to one dog year of life. Like everything else, if only it were that simple.

How "Old" Is My Dog? Telling Time on a New Clock

People have been comparing human-to-dog years for centuries. A floor inscription made in 1268 by the artisans in Westminster Abbey made a prediction for Judgment Day: "If the reader wisely considers all that is laid down, he will find here the end of the primum mobile;

a hedge lives for three years, add dogs and horses and men, stags and ravens, eagles, enormous whales, the world: each one following triples the years of the one before." By this math, a human lives to eighty and a dog lives to nine. Luckily for both humans and dogs, chances are we can live longer than that if we make the wise choices.

The seven human years for every dog year equation could have originated from a general observation that people on average live to around seventy years and dogs about ten. But some experts think it could have just been a way to encourage dog owners to bring their pets into the vet at least once a year, under the thinking that dogs age much faster than we do. Which is true, but a more accurate way to draw parallels between the two species is to use the following formula: The first year of a medium-size dog's life is equal to about fifteen human years; the second year of a dog's life is equal to about nine years; and each year thereafter is equal to about five human years.

Most charts that show a dog's age also factor size into the equation. Charts like these, however, are constantly being challenged and revised to reflect emerging science. Other studies suggest one-year-old dogs have a "human age" of about thirty, while by the age of four, they are about fifty-four in human years; by fourteen, they are on par with humans in their mid-seventies. But therein lies the rub: What defines "human years" (and, by extension, why do we feel the need to arbitrarily pin or parallel dog years to "human years")?

Calculating a dog's age—or anyone's age, for that matter—should also differentiate between chronological age versus biological age. We all know people who seemingly defy their age. The seventy-year-old who looks and acts ten years younger. Or the nine-year-old German Shorthaired Pointer that looks and acts like she's four. The concept of age is relative—relative to how our bodies are functioning, how well we take care of ourselves, and how we behave in the world. Years ago you might have heard of a special calculator called the RealAge test, founded by Dr. Michael Roizen, who is currently the chief wellness officer at the Cleveland Clinic. The test aims (unscientifically) to predict longevity based on things like how much exercise we get, whether we smoke, what kind of diet we follow, what our medical lab work (e.g., cholesterol, blood pressure, weight) looks like, and our medical history. Clearly, no test can actually predict life expectancy, but these tests are good at showing where we could be putting effort into our health equation.

Scientists have tried to come up with novel, data-driven ways to measure true biological age, and various methods have emerged over the years. Again, no single test can be totally accurate, but they are nonetheless fascinating and worth investigating. The length of our telomeres, for one example, has been purported to indicate how well we are aging. Telomeres are caps of DNA at the ends of chromosomes that protect our cells from aging. Telomeres naturally shorten over time, so it stands to reason that having shorter telomeres sooner in life is not a good sign. One of the more interesting, cutting-edge ways of measuring age, and which brings us back to the difference between dogs and humans in the aging process, is the so-called epigenetic clock. Pioneered by geneticist Dr. Steve Horvath at UCLA, the epigenetic clock relies on the body's epigenome, which comprises chemical modifications that tag DNA. The pattern of these tags changes during the course of life and tracks a person's biological age, which can lag behind or exceed chronological age.

These epigenetic marks are important for our health and longevity and also for how these traits are passed on to future generations. Even the foods you eat, the air your dog breathes, and the stress you both experience influence your DNA (hard drive) through the epigenome—the software that determines which genes will be active.

This is hugely important for the people who pay lots of money for a genetically stellar dog that is regularly exposed to environmental substances that negatively affect the epigenome; the dog's health can unravel due to epigenetic triggers. Likewise, dogs that are found to have genetic variants, or a genetic predisposition for a disease, will not automatically manifest unexpressed DNA and get the disease; you can have a dramatic, positive impact on the epigenome. Your dog could be a genetic mess and never express any signs of disease.

Taken together, this makes clear that lifestyle and environment are manipulating your dog's epigenome. It is our responsibility to create healthy surroundings to influence our pet's DNA.

Influencing the epigenome is a powerful piece of knowledge for guardians. It's empowering to know that even if my dog is inbred or has massive genetic flaws, there's still lots of hope of radically improving quality of life and slowing disease progression, thanks to epigenetics. Regardless, addressing all known epigenomic influences is the only way to slow aging and, by default, create the potential for exceptionally long-lived animals. Although it's beyond the scope of this book to address epigenetic issues at breed-specific levels, you will gain the upper hand in supporting positive epigenetic expression if you implement our recommended strategies. (For more personalized ideas, please go to www.foreverdog.com.)

TOP EPIGENOME TRIGGERS

Nutrient levels in food

Polyphenol levels in food

Chemicals in food

Physical activity

Stress

Obesity

Pesticides

Metals

Endocrine-disrupting chemicals

Particulate matter (secondhand smoke)

Air pollutants

Taking into consideration the changes made to human and dog DNA over time, in 2019, researchers at the University of California, San Diego, suggested a new clock. The developmental pattern of a dog remains the same no the matter the breed: they reach puberty around ten month and die before hitting twenty. In order to optimize their chances of finding a genetic link to aging, these researchers settled on one breed: Labrador Retrievers.

They gathered a pool of 104 dogs between the ages of four weeks to sixteen years old in order to study DNA methylation patterns in the genomes. They found that Labrador Retrievers and humans have comparable age-related methylation patterns. They also found that some of the genes that take part in development also methylated during aging in both humans and Labs. This suggests that some portion of the aging and development process overlap rather than being two separate processes. Furthermore, it seems these processes and changes are evolutionarily conserved in mammals.

These researchers developed a new way to calculate "dog years," and it is slightly more complicated than the old-fashioned method, which was simply "multiply by seven." For dogs over the age of one, the formula says that a canine's human age roughly equals 16 x ln (dog's age) + 31. To work out your dog's "human" age, first enter the dog's age then press "ln" on a scientific calculator. Then multiply the figure you get by 16, and finally add 31.

Our life stages match with dogs. Both a seven-week-old puppy and a nine-month-old human are developing teeth, making them equivalent in age. The expected lifespan of a Labrador Retriever also matches with a human's, being twelve and seventy, respectively. In general, your dog's aging clock is swifter in the beginning compared to a human's. For example, a two-year-old Lab might still behave like a puppy, but she is actually in the asymptomatic aging process, and then slows down from there.

As you can imagine, most dog lovers are not happy about this discovery, but these research results certainly spurred on the popularity of telomere measurements as a blood-testing trend in the human biohacking space, and now there's even one lab offering telomere measurements for dogs.

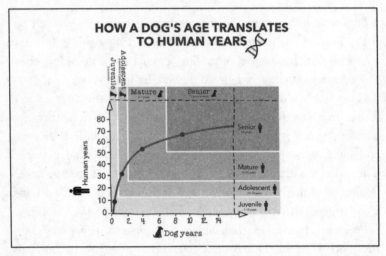

This chart shows the difference between how dogs versus humans age over time. It's based on some pretty complex calculations and several studies, notably one led by Dr. Trey Ideker at UC–San Diego, from which this image is adapted. The shaded, outlined boxes indicate the approximate age ranges of major life stages based on common aging physiology. Juvenile refers to the period after infancy and before puberty (2–6 months in dogs, 1–12 years in humans); adolescent refers to the period from puberty to completion of growth (6 months to 2 years in dogs, approximately 12–25 years in humans); mature refers to the period from 2–7 years in dogs and 25–50 years in humans; senior refers to the subsequent period until life expectancy, 12 years in dogs, 70 years in humans. Dog life stages are based on veterinary guides and mortality data for dogs. Human life stages are based on literature summarizing life cycle and lifetime expectancy.

Going solely on this formula, your dog's "human" age may not line up perfectly. This formula may not include necessary variables that skew the aging process like breed and size. However, this formula still carries more weight than the "multiply by seven" dog years myth.

AMYLOID AND AGING

Most of us recognize that with advanced age often comes stiffness, arthritis, and joint issues. We see a lot of aging dogs walking around very stiff, like they have bamboo legs. The same degenerative changes we see in our dogs' outward appearance can also be happening in their brain. Now we are starting to understand the relationship between amyloid formation and aging. You might already be familiar with beta-amyloid protein in the brain that, when misfolded, adversely accumulates to create sticky globs called "plaque," can be a hallmark sign of Alzheimer's disease. Dogs, too, develop Alzheimer's disease–like beta-amyloid, which is associated with cognitive decline. This is why scientists study dogs—to further their knowledge of Alzheimer's disease and search for treatments. The relationship between brain health and cardiovascular health rings true for them as it does for us. Stiffening of the arteries appears to be linked with the progressive buildup of this beta-amyloid plaque in the brains of elderly patients—even among those without dementia. Such a finding points to a relationship between the severity of vascular disease and the plaque that is a hallmark of neurodegenerative disease. Which means the key to saving your brain and your mobility (no bamboo legs!) is to preserve cardio fitness. What's good for heart health is good for brain health, whether you're a dog or human.

The process of methylation continuously repairs DNA. The adding or subtracting of a methyl group causes profound biochemical changes to occur in our body, as well as in our dogs, as they activate and deactivate the body's life code and affect core life processes. So when methylation goes awry or becomes imbalanced, trouble can arise. Methylation defects have been associated with cardiovascular disease, cognitive decline, depression, and cancer, with similar studies being conducted on dogs. Plenty of questions remain. Are alterations in methylation a cause or an effect of aging? Or perhaps they're linked to aging in some other way. "No one knows, it's all speculation," according to Dr. Trey Ideker, a professor of genetics at UCSD who led the methylation study. In 2020, when Augie, the Golden Retriever in Tennessee, broke the record as the world's oldest Golden at age twenty, Dr. Ideker's insights reverberated around the planet when people sought the secrets to her longevity. Now the goal is to figure out what sets the rate of methylation, and why it happens faster in some animals than others. By understanding this genetic clock, we may be able to gain an upper hand in controlling the aging process—in our pets and in ourselves.

The Super Science of DNA Variations

Single nucleotide polymorphisms (SNPs) represent variations in DNA sequences. SNPs are alterations in the set of genetic instructions that are thought to provide the genetic markers for a response to disease, environmental factors (including food), and drugs. These variations, which are specific edits in the DNA code, can translate to things like a trait for color coat, or a higher susceptibility for developing cancer, or the inability to clear histamine in dogs (and people).

Some SNPs and combinations of gene variants can profoundly impact a body's ability to make and use different nutrients that are critical for things like reducing inflammation, promoting normal detoxification and immune function, and producing healthy neurotransmitters. Certain genetic variants can result in the body's receiving different or incorrect cellular instructions. For instance, the selection of an alternative amino acid during protein synthesis will change the shape of the resulting protein. This means that what happens downstream in the body will be different or impact the function of other cells, organs, and tissues—all because of genetics. But what if there's not enough bioavailable amino acids in the diet for the first, second, or third amino acid choice? This is where nutrition impacts DNA. We'll talk about the quality and quantity of amino acids (proteins) and all the other important nutrients not found in highly digestible quantities in ultra-processed dog food later, but genetic variants plus poor-quality nutrition are part of the reason we believe so many dogs are diagnosed with degenerative diseases at midlife, or before age ten.

It's important to realize that these DNA differences do not necessarily *cause* a disease, but they can be markers of the relative risk of disease. By the same token, if your dog has no known genetic risk markers for a condition, it doesn't guarantee that she won't develop it, but it does mean that her risk is lower than for dogs with those particular risk markers. Since the completion of the Human Genome Project, thousands of studies have been published that describe the associations between SNPs and hundreds of specific diseases, traits, and conditions. The same type of study is underway for dogs. Because food speaks to our genome, scientists are finding SNP patterns that affect methylation pathways in people and dogs that can inform better nutrition for overall wellness; this new frontier is referred to as methylgenetic nutrition.

A dog's DNA controls his physiology, including how well his body makes nutrients and enzymes and clears toxic substances. If you or your dog has a wad of variants (SNPs) preventing normal physiology, metabolism, and detox mechanisms from functioning optimally, it's easy to see how the body can break down without intervening with key nutritional support. Research is finding there are genetic variants that impact dog behavior as well, including fear.

The great news about the speed at which the human medical community is responding to genetic diagnostics is that personalized methylgenetic nutrition and functional genomic nutritional analysis are already growing trends for biohackers, athletes, and people looking to optimize their health through customized nutrition and supplementation. A simple DNA saliva test reveals a person's unique genetic variants, and this raw data can be uploaded into specialized software, along with the patient's lab work, which allows doctors and nutritionists to identify metabolic pathways that need extra support, completely customized around a patient's unique genetic profile. Recommendations can then be made for providing the missing cofactors or nutrients in a form the individual can metabolize.

Talk about personalized medicine and nutrition! Our genetic makeup can guide us toward which medicine, chemotherapy treatments, vitamins, minerals, and supplements work best for our health—and which to avoid. Right now, we test only for genetic disease markers in dogs, but veterinary medicine thankfully is also headed in the direction of personalized medicine and nutrition. A few years down the road, we will see genomic nutrition analysis available for veterinarians, and more species will benefit from customized nutrition and supplementation, as well as medical and drug protocols based around our animals' unique genetic makeups. There is already a burgeoning sector of the wellness industry tailoring nutraceutical products to dogs based on their breed, DNA, lifestyle, and age.

LONGEVITY JUNKIE TAKEAWAYS

➤ Life is an ongoing cycle of continual destruction and rebuilding. Aging is also a normal, ongoing process that involves multiple actions in the body that reflect both genetic and environmental forces.

➤ The various pathways, or hallmarks, of aging—and the myriad roads to cellular, organ, and systemic dysfunction— are the same for both humans and dogs. Although dogs go through similar stages of development to those of humans, they age much faster and thus provide an opportunity for us to study them for clues to programming optimal aging.

➤ Genetic mutations and/or nutrient deficiencies can up or down regulate a dog's rate of methylation, which in turn sets the stage for accelerated or slowed aging.

➤ Size matters: Larger dogs are more likely to die prematurely than smaller dogs, and this is partly due to discrepancies in metabolism and weight-related risks for degenerative diseases. Age and breed are dogs' most monumental risk factors.

➤ Spaying or neutering your dog too early can have long-term effects on their health and behavior. When desexing puppies before puberty, consider a hysterectomy or vasectomy instead of full removal of sex organs.

➤ Many dog genome projects are underway to map certain canine genes at risk for disease.

➤ Adverse chemical reactions occur when glucose (sugar) and protein react with each other, both inside a warm body and when food is mixed together and heated. The result is harmful products like MRPs (including AGEs and ALEs) that wreak biological havoc. These compounds, combined with other insulting ingredients such as mycotoxins, glyphosate, and heavy metals, further adulterate commercial pet food.

➤ A simple way to gauge the healthfulness of your dog's pet food is to count how many times it's been cooked/heated in its processing (we'll cover the details in Part III).

➤ Although research, inspection processes, and regulation differ among the human and pet food industries, both use the same slick marketing ploys and both continue the unabated push for ever-more processed foods.

➤ DNA is static, but how it behaves or expresses itself is highly dynamic—a phenomenon of epigenetic switches. Among the top epigenetic triggers that have the power to change how DNA acts are nutrients in food, environmental insults, and lack of physical activity.

➤ Autophagy is an important biological process that helps keep a body clean and tidy. There are times when you want to activate autophagy, and you can do so through diet, meal timing, and exercise.

➤ Like children, dogs need the right home environment, with daily guidance on behavior and sociality, to mature into well-behaved, fun-loving fur balls that handle stress well.

➤ The age of dogs relative to humans is not merely a factor of seven. There are a few ways to gauge a dog's "age," but the fortitude and agility of a dog's underlying epigenetic switches are the most important factors in determining wellness or dysfunction.

PART II

Secrets from the World's Oldest Dogs

The Long Tale

4

De-Aging through Diet

How Food Is Information for Health and Longevity Genes

> What you eat literally becomes you; you
> have a choice in what you're made of.
>
> —Anonymous

> Everyone has a physician inside him or her; we
> just have to help it in its work. The natural healing
> force within each one of us is the greatest force
> in getting well. Our food should be our medicine.
> Our medicine should be our food. But to eat
> when you are sick is to feed your sickness.
>
> —Hippocrates

In 1910, an Australian cattle dog named Bluey was born in Victoria. He would go on to live for twenty-nine years and five months, making the Guinness World Records as the planet's oldest dog. He lived on a farm where he worked among the sheep and cattle. In 2016, an Australian kelpie named Maggie passed away in her sleep; she was reportedly thirty years old. Like Bluey, Maggie lived on a farm, but she couldn't claim the prize for being the longest-lived dog because her owner lost the paperwork to verify her age. These two remarkable dogs had a lot in common. They spent their days running around outside in wide-open spaces so they got plenty of exercise and exposure to nature. Check, check. Maggie's owner, Brian, said she ran miles every day from just following him on his tractor. Life on the farm meant access to a lot of fresh, whole foods, too. Maggie's balanced diet was mostly raw foods that included a good amount of protein and higher fat intake, devoid of processed dog foods. Check. And their lifestyles in their respective settings were low-stress, high-quality. Check.

Both Bluey and Maggie are known in dog circles as Methuselah dogs. These exceptionally long-lived outliers are making headlines around the world, thanks to Dr. Enikő Kubinyi and her dog-centric research team from Budapest. Methuselah was a biblical patriarch and a figure in Judaism, Christianity, and Islam. He is said to have lived to the age of 969, the oldest figure mentioned in the Bible. In human studies, centenarians (one hundred plus years of age) typically are called Methuselahs. In the canine world, dogs can be considered Methuselahs if they live seventeen or more years (but as we've covered, there is considerable variation due to different factors such as breed and size). Mixed breed dogs who live at least twenty-two and a half or more years are considered to be Methuselah dogs. Studies show that only one out of a thousand dogs lives up to twenty-two to twenty-five years. The vast majority of dogs die younger from various diseases, many of which are outcomes of modifiable risk factors like diet and exercise.

It bears repeating that dogs are increasingly being used in research circles to study the aging process, because they are great models for helping us understand different aging pathways and lifestyle modifications that can extend the quality and quantity of life. The research is symbiotic—as dogs clue us into the aging process, we simultaneously learn how to extend their lives as well.

We are often asked: What's the one thing we can do to help our pets live longer? You might be surprised to learn that the same "one thing" applies to us: optimize diet. Put another way, **eat better, eat less, and eat much less often.** While simple in theory, this is not always easy to implement in real life. We are endlessly distracted by a cornucopia of shiny, brightly packaged options, many of which we'd do well to avoid. Think about your own dietary habits. How many trendy diets have you tried either to lose weight or to deal with a chronic condition? Whether you've ever counted carbs and calories, debated whether or not to have seconds or thirds at the dinner table, or followed a strict, by-the-book protocol and tracked your meals over days, we're willing to bet that at least once in your life you've consciously made an effort to clean up your diet. Most of us do at some point, and then we fall off the wagon until New Year's.

Unlike us, our dogs don't have the power of dietary choice. They are wholly dependent on us—their caretakers—to make the right choices *for* them. Put yourself in your dog's shoes (paws): You're now fed every meal by your beloved human. You have no say in the matter. Primal hunger drives you to eat whatever is put in front of you. If all you could eat was highly processed food, how long would it take for you to feel its effects on your body (your waistline), not to mention your mind or immune system? Probably not long—days to weeks. The weight would creep up. You'd feel sluggish and mentally foggy. Restful sleep would be impossible; your overall stress levels and anxiety would rise, manifested in elevated cortisol levels. And you'll eventually crave a clean meal of whole, fresh, unprocessed foods directly from nature. That's how our primate ancestors ate, so needing nutrients from a wide array of whole fresh foods is already programmed into our genome. For a dog's ancestor, the wolf, finding food was a chore that entailed strategy and smarts. It also involved lots of physical movement.

Both ancestral wolves and modern wolves are classic carnivores. They prefer to eat large, hoofed mammals such as deer, elk, bison, and moose. They also hunt smaller mammals such as beavers, rodents, and hares. Their diet is primarily protein and fat, unadulterated by processing. Likewise, ancestral humans (who lacked the benefits of the modern food industry and global supply chain) hunted and gathered; they also consumed a diet rich in wild game, fish, and edible plants, including nuts and seeds. Wolves also ate berries, grasses, seeds, and nuts, making both species' dietary habits vulnerable to the omnivore argument. Early humans were lucky to come across sweet fruit when it was in season, and studies show that those natural and ancient fruits were probably tart and bitter; we've bred ours over the past two hundred years to be extremely sweet and sugary. And large, ancient apples bore almost no resemblance to today's apples. Just as we bred dogs to suit our fanciful desires, we also bred fruit to suit our preferences; today's fruit has become something akin to bulk candy plus a mediocre multivitamin.

Our ancestors also didn't eat as much or as frequently as we do today. They had to work hard (i.e., exercise) to get their food, and breakfast likely was not the most important meal of the day because it took the entire day—or several days—to find food. Sometimes days would go by without a bite, but that was okay because the human body had evolved to endure periods of starvation. It had to for survival. We have built-in biotechnology to handle long stretches of food scarcity. The problem arises when our Stone Age biotechnology meets the modern era of convenience and processing. The same can be said for our dogs who "grew up" with us in the Stone Age but live now in the twenty-first century. Dogs still fit the definition of a carnivore, like house cats. They have an ultrashort gastrointestinal tract, they are unable to synthesize vitamin D from sunlight, and they lack salivary amylase (the enzyme that digests carbs). Because domestic dogs have increased their *pancreatic* secretion of amylase, however, many veterinarians assume dogs can be vegan. We disagree.

A calorie is not a calorie, according to diet and obesity researcher Dr. Jason Fung, who has written extensively on the power of time-restricted feeding and the difference between a calorie from, say, a stack of syrupy pancakes versus a calorie from a veggie omelet. His insights, now widely accepted in the human nutrition community, fly in the face of almost every recommendation of today's board-certified veterinary nutritionists. Metabolically, a diet of all carbs acts completely differently than a diet composed of balanced protein and healthy fats, and not all carbs are created equal—how your body responds to a carb-rich meal depends on your chemical makeup and whether you digest those carbs quickly or slowly. An individual digesting a bowl of slow-releasing carbs like grilled veggies will feel different from someone digesting a bowl of fast-releasing carbs like cereal. Those slow-releasing (or "slow-burning") carbs will keep you feeling fuller longer than the ones that rapidly get digested ("fast-burning") and keep you craving more and more. One thing all vets can agree on is that healthy dogs (unlike cats) are evolutionarily adapted for starvation. Dogs were not all that successful at hunting at dawn and dusk every day. They could go for days without catching any food and would eat what they could find—carrion, vegetation, acorns, berries—if they were unsuccessful in hunting for a fresh dinner.

Cancer researcher Dr. Thomas Seyfried has been teaching neurogenetics and neurochemistry as it relates to cancer treatment at Yale University and Boston College for more than thirty years. He told us about a dog named Oscar who—back in the day before ethics boards oversaw the humane treatment of research animals—was fasted for more than one hundred days at the University of Illinois veterinary school. According to Dr. Seyfried, Oscar was sent back to the farm and was still able to jump three feet over a fence and into his kennel. The point of mentioning this gross study here is that healthy dogs *can* successfully fast. We don't recommend specific fasting protocols in this book because each protocol should be tailored around the age and health of each animal. But we will introduce you to fasting-mimicking strategies that allow your dog to harness the benefits of fasting but without having to skip meals.

If you could time-travel—with your dog and a bag of groceries in hand—to a group of primitive humans gathered around a roaring fire to share a fresh catch, they'd be stunned, perhaps speechless, at your goods. They may not recognize anything you pull from your bag, even if you denude the food from its twenty-first-century packaging. The nutrition labels alone would confuse them. They would not recognize the ingredients (let's assume they can read). And the dog food? Let's say you packed a bag of typical kibble and canned dog food in your time-traveling suitcase for purposes of this experiment. The companion dogs around your ancestral comrades wouldn't recognize either the kibble or the canned muck and probably wouldn't touch any of it (they'd be sharing the scraps tossed to them by their masters). Meanwhile, you're sitting there eating a meal that may as well have come from outer space. And your twenty-first-century dog would be decidedly envious of his predecessors, salivating over what his fellow counterparts from another era get to devour in their family units.

The Power of Food

Understanding the power of food is essential to gaining better health and extending healthy life for you and your dog. Food is the cornerstone of lifestyle medicine. As we've been saying, food is much more than fuel for your body; food is *information* (literally: "it puts the *form into* your body"). It is simplistic and misguided to think of food merely as calories for energy (fuel), or that food is just a bunch of micronutrients and macronutrients (building blocks). To the contrary, food is a tool for epigenetic expression; your diet and genome interact. In other words, the food you eat speaks to your cells, and that critical communication instructs your DNA's functionality. **Because of its continuous and life-long impact, nutrition might be the most important environmental factor for health**. Indeed, food is one of the most potent and influential ways to build or destroy health in our companions; it can either heal or harm. Molecular nutrition research strives to understand this interaction. Nutrigenomics (sometimes referred to as nutrigenetics), is the study of how nutrients and genes prevent, treat, and manage diseases, and understanding it is integral to helping your Forever Dog.

Nutrition is not extensively taught in medical and veterinary schools, at least not in the ways other subjects—physiology, histology, microbiology, and pathology—are covered. Don't misunderstand: Those are all important subject matters. But there is no "ology" for nutrition, and it gets the short shrift in veterinary training. Times may be changing in this department for future generations of doctors and vets because we're all finally waking up to the dearth, but by and large medical schools and vet schools remain steeped in traditional methods: we learn the basics about biology and physiology and then we learn how to diagnose and treat disease. There is very little teaching and learning about *preventing disease in the first place.* As with human medicine, veterinary medicine remains stuck in the age-old paradigm of managing disease and controlling symptoms rather than preventing its onset in the first instance. Your vet isn't purposely withholding this critically important information; rather, she doesn't discuss targeted nutritional interventions, lifestyle choices, and risks and preventive strategies with you because they didn't teach her these things in vet school.

Moreover, the nutritional education that veterinary students *do* receive—much as with medical students—may be biased because courses are commonly taught by nutritionists endowed by commercial pet food conglomerates. Vets get their information largely from within the processed pet food industrial complex—the manufacturers of the very foods that contribute to poor animal health. Talk about the fox guarding the henhouse! Frankly, it's even worse: The fox is *inside* the henhouse!

Nutrition illiteracy is a global problem. A 2016 survey of deans and faculty members from sixty-three European veterinary colleges revealed 97 percent of respondents believe the ability to perform a nutritional assessment on patients (animals) is a core competency.

But only 41 percent of the respondents reported satisfaction with the skills and performance of their school's graduates in veterinary nutrition.

The number of fresh-food-literate vets is growing rapidly, but not because vet schools are teaching students how to calculate nutrient requirements for homemade, nutritionally optimal recipes, and not because small-animal nutrition classes are exploring how pet food processing techniques (extruded, canned, baked, dehydrated, freeze-dried, gently cooked, and raw) affect nutrient losses. This demand is *consumer driven*. Pet owners are insisting on fresher foods, and veterinarians are faced with either educating themselves in order to support their clients or losing them. Because many people can't yet have transparent and instructive conversations with their vets about how to feed nutritionally balanced homemade diets, online directories such as www.freshfoodconsultants.org are filling the void, providing nutritionally complete recipes to pet owners around the world.

When the Planet Paws Facebook page began innocently enough in 2012, an image and list of all the ingredients found in a typical bag of commercial dog food garnered half a million shares overnight. The audience grew quickly, bringing to light a voracious hunger for knowledge; people were desperate for knowledge about the proper way to feed and care for their pets. One of the most popular posts about rawhide chews has been read more than half a billion times, and the accompanying video describing how rawhide is made has been watched over 45 million times. By 2020, Rodney's Planet Paws page had nearly 3.5 million followers, and the conversations most likely to go viral are diet-related.

Clearly, pet owners are desperate for guidance. They want fact-driven, science-backed wisdom about pet nutrition and wellness. No gimmicks, no false advertising. It's no wonder one of the most-viewed TED Talks in history on the subject of dogs came from one of us (Rodney), and the other (Dr. Becker) became the first veterinarian in the world to present a TED Talk on species-appropriate nutrition. We love watching the pet food industry evolve before our eyes as more transparent, ethical pet food companies emerge to crowd out highly modified options that diminish the health and longevity of Forever Dogs. A revolution is afoot. If you haven't begun to participate, you will by the end of this book (if not by the end of this chapter!). And we are not alone in this thinking. According to food scholar Marion Nestle, "We are in the midst of a food revolution." This is true for both humans and our furry counterparts, and she calls this the "good pet food movement." Just as humans may add organic, fresh, locally sourced foods into their diets, the same can be done for dogs.

Imagine getting *optimal* nutrition from one bag of food your entire life. Sound impossible? It is. You can't even rely on a single protein shake to supply all your nutritional needs. As we mentioned in Part I, people who subsist on nutritional beverages billed as all-in-one drinks, with vitamins and ingredients to fill in nutritional gaps, do so only for brief periods of time and for particular reasons (e.g., during hospitalization). These drinks, which qualify as processed foods, are not meant to be a sole food source for a lifetime even though they contain all the recommended daily nutrients. No one—not you or your dog—can thrive on a monochrome, processed diet.

People are waking up to the fact that animals need more than pellets to nourish their bodies. A 2020 study found that only 13 percent of pet owners feed ultra-processed food exclusively. This is great news because that means 87 percent of pet owners are adding other foods to their pets' bowls. Some countries are further ahead in the race to recover companion animal health, with Australians topping the list of pet owners feeding more fresh food than canned or kibble.

Part of the reason fresher food has created such a divide between veterinarians and pet parents is because it's really the only pet food choice that offers a DIY option (you can't make kibble at home), and therefore there's the potential to do it wrong. And people do.

Loving pet owners with great intentions have created some un-intentional nutritional disasters by guessing at what constitutes nutritionally balanced meals for their pets. Every veterinarian we know can relay at least one story about the unfortunate con-sequences of feeding (inappropriate) fresh food to their dog, from acute diarrhea (from changing foods too quickly) to fatal nutritional secondary hyperparathyroidism (a metabolic bone disease caused by months to years of inadequate calcium ratios). There are endless horror stories about all the things that can go wrong with nutri-tionally unbalanced homemade meals, which is one reason the com-mercial nutritionally complete raw pet food industry has boomed in the last fifteen years. In fact, **the fresh pet food category is one of the fastest-growing segments in the pet food industry**, much to the Big Five's dismay. Five massive companies dominate the $80 billion ultra-processed pet food industry, and none of them currently produce a fresh pet food made with quality, human-grade ingredients. (Mars Petcare, which clinches the number one spot by a long shot, now includes three of the top five pet food brands in the world—Pedigree, Whiskas, and Royal Canin; the behemoth is a growing segment of approximately fifty brands.) The more profits these companies lose to the myriad human-grade, fresh pet food companies popping up in droves, the more fearmongering about fresh food diets we read from these companies. We'll get into the details of how to make sure the commercial dog food you buy is nu-tritionally adequate later on, but when it comes to homemade diets, they can be the best or worst food you offer your dog, depending on its nutritional adequacy (or inadequacy).

Homemade diet disasters are the result of well-intended but un-informed humans guessing at what to do. That's another reason we wrote this book: to give you a solid blueprint to follow, backed by science. Interestingly, you rarely read about the tens of thou-sands of pet parents who follow nutritionally complete recipes, do things right, and restore their pet's health on their own. One rea-son for the rapidly growing number of fresh-food-literate veterinarians around the world is the sheer number of near-miraculous recoveries in otherwise hopeless cases. It's difficult to ignore the results, and many veterinarians have come around after witnessing numerous canine patients enjoying improved health or disease regression after their owners instituted a fresh food diet. Most likely, for every horror story your vet shares about nutritional secondary hyperparathyroidism, she can also (possibly begrudgingly) recall the dozens, maybe hundreds of clients choosing this "alternative feeding style" she does not condone . . . and having life-changing results. Board-certified veterinary nutritionist Dr. Donna Raditic suggests that veterinarians who recommend only ultra-processed pet food as a nutritional standard of care may be eroding pet owner confidence in veterinarians. The 2020 study polling 3,673 pet owners from Australia, Canada, New Zealand, the UK, and the United States found that 64 percent of owners offered homemade meals to their dogs. We venture to guess that most of these owners do not discuss these feeding choices with their dogs' doctor to avoid confrontation.

It makes us wonder how many success cases veterinary skeptics must see (without one's own involvement as a doctor) before curiosity sets in, or at least a softening toward having open conversations with their fresh-feeding clients. The good news is that millions of animal guardians have become empowered, knowledgeable advocates, dramatically improving their dog's nourishment, wellness lifestyle, and environment. As a result, they have positively influenced the state of disease in their own animals. We see veterinarians with a growth mindset be most inquisitive about emerging animal health trends. This has resulted in thousands of veterinarians worldwide investigating what's behind these dramatic recoveries in which they played no role. As with any health paradigm shift, a rift begins to form between the old and the new. The Raw Feeding Veterinary Society, a professional organization for vets wanting to learn more about fresh food, is just one upshot to the massive pressure for change pet parents are putting on the industry.

Dozens of smaller, independent, fresh pet food companies have sprung up around the globe, with passionate, board-certified nutritionists leading the charge in producing real-food meals made with fresh ingredients that are minimally adulterated. Of course, these companies comprise a tiny sliver of the overall pet food market, but they are poised to grow and expand their presence as pet owners like you become more informed about the risk of a lifetime of fast-food consumption. It's a welcome change after decades of criticism, directed at the fresh-feeding community of vets and health-conscious pet lovers, for suggesting pets should be transitioned to less-processed foods. And, frankly, the word "raw" is easily misconstrued and misunderstood. We rarely use it due to the stereotype that raw food means spoiled meat instead of meat locally sourced from a nearby butcher. Additionally, it is only one option for "fresher" pet food. However, the poor connotation around the word "raw" has steered pet parents away from the type of food their companions naturally crave and are genetically predisposed for. In Part III, you'll learn there are a variety of pet food categories within the "fresher food" category in addition to raw food. And within the raw food space there are a half dozen more choices, including pasteurized raw food that's been sterilized.

Let's not forget that we were all once told trans fats (heavy margarine) were best, and that doctors once helped Big Tobacco sell cigarettes. (No joke: In 1946, Reynolds launched an ad campaign with the slogan, "More doctors smoke Camels than any other cigarette.") When you let the emerging science speak, you can't argue. And if you aren't ready to feed raw food, no problem. You can make or buy gently cooked fresh food; the remarkable health benefits of reducing the amount of ultra-processed food consumption will positively influence your dog's health.

We get it: Just as we're not going to feast on a plate of raw chicken and a glass of untreated water with a side of oysters from a polluted bay, we're not going to feed our dogs anything that has the potential to sicken them. This is about using wholesome ingredients that are safe, delicious, and nutrient dense in order to stack the deck in your dog's longevity favor. Dogs may enjoy a lot of foods we do, but we must respect species differences, and we'll show you how to do that. And also for the record: We're not suggesting you feed your dog like she's a wolf. Virtually all bags of dog food feature wolves on their packaging, which is reminiscent of the slick images of happy, exuberant, healthy-looking people on the packaging for über-processed human foods. We all know those foods don't support wellness in the long run. But brilliant marketing is designed to target and exploit our vulnerabilities, while titillating our taste buds. Our dogs hunger for healthy protein and unadulterated, healthy fats with few carbs. That's exactly what their ancestors selected, and it's decidedly *not* what's found in the glossy wolf-plastered bags of dog food.

Raw, unadulterated food is what dogs evolved eating, and they certainly haven't lost this evolutionary adaption in the last hundred years. But that's not the only option if your goal is to avoid the ultra-processed category. Obviously, our goal is to minimize the amount of highly refined foods our family consumes, but not all processing techniques are created equally. In Part III, we'll teach you how to evaluate dog food brands with some easy criteria— adulteration math—that makes it simple to see what category dog food products fall into. The industry knows you're looking for less-processed pet foods, so these terms are tricky. The pet food industry appropriated the terms "natural," "fresh," and "raw." And now ultra-processed pet food companies have hijacked the term "minimally processed," slapping it on their bags of kibble and completely deceiving most pet parents.

Determining the minimally processed, processed, or ultra-processed status of foods can be difficult. The truth is, unless your dog is catching prey or eating blackberries directly from the garden, all commercial diets you serve him have been processed in some way. In theory, washing and chopping just-picked veggies is

a process, but we're talking about the more widely accepted definitions by the nutrition community:

Unprocessed (raw) or "fresh, flash-processed foods": Fresh,
raw ingredients that are slightly altered for the purpose of
preservation with minimal nutrient loss. Examples of minimal
processing techniques are grinding, refrigerating, fermenting,
freezing, dehydrating, vacuum packaging, and the pasteurizing of
food (according to NOVA food classification system).

"Processed foods": The previous category's definition
("minimally processed foods"), as modified by an additional
thermal (heat) process. (So there are two processing steps,
including heat adulteration.)

"Ultra-processed foods": Industrial food creations (which can't
be replicated at home) that contain ingredients not found in
home cooking; require several processing steps using multiple
previously processed ingredients, with additives to enhance taste,
texture, color, and flavor; and are produced through baking,
smoking, canning, and extrusion. Extrusion refers to forcing
a pumpable product or mixed ingredients—in this case, dog
food—through a small opening to shape materials into a designed
fashion. Extrusion was developed in the 1930s for dry pasta and
breakfast cereal pellet production. Then, in the 1950s, it was
applied to pet food manufacturing.

This makes things crystal clear: Fresher, flash-processed dog
food contains ingredients that have undergone *one adulteration*. This
includes raw (frozen) dog food diets, high-pressure-pasteurized
(HPP) raw dog food diets, freeze-dried, and dehydrated dog foods
made from raw, previously unprocessed ingredients. They're called
"flash-processed" because the adulteration or processing insult happens in a flash—a very short period of time—and only once. In theory, this category of pet food could be called "ultra-*un*processed," as

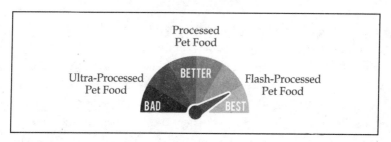

it's on the opposite end of the processing spectrum, as compared to ultra-processed pet food.

Processed dog food has undergone an additional thermal (heat) adulteration step or contains ingredients that did. This category includes gently cooked dog food diets and freeze-dried or dehydrated diets made with ingredients that were previously processed (not raw). This category of food is healthier than ultra-processed pet foods because the ingredients have not been refined or heated repeatedly.

Ultra-processed dog food has multiple heat adulterations, includes previously heat-refined ingredients, and also contains industrial additives not available to the consumer. For instance, we can't buy high fructose corn syrup or chicken meal (or any meat meal) in the grocery store; only the food industry (human and pet, respectively) has access to these ingredients. Putrescine and cadaverine (trust us, you don't want to know any more about these two flavor enhancers, their names say it all) are just two of the many additives companies use to entice dogs to eat dry food. They are not available to consumers. Corn gluten meal, a common ingredient in many veterinary lines of pet food, is available to consumers only as a weed killer sold in home and garden stores (not grocers). And there's no home version of a kibble maker. By definition, ultra-processed dog foods include most "air-dried" dry dog diets and some dehydrated dog diets (those not made from fresh, raw ingredients), and all canned, baked, and extruded dry dog foods. In Part III, we'll give you some simple tips to decipher all the confusion. This information will be shocking to some and infuriating to others, but it is important information to know for the health of your pet.

One of the most interesting and recurring experiences we had while interviewing some of the world's top longevity experts has been their notable responses when they connected the dots on this subject. More often than not, after powerful interviews with brilliant scientists talking about how their research demonstrates food healing or harming, speaking to the epigenome, or smashing/saving the gut biome, we asked what they feed their own dogs. As those conversations unfolded, we heard several "Oh mys" and "I never thought about how these discoveries impact other mammals" and lots of "Please send me the book when it's done, but tell me what to do now!" reactions. We get it: It's overwhelming to learn we've inadvertently let the massive worldwide fast-food industry dictate what constitutes "healthy food" and snacks, for both us and our pets.

Doctors are urging us to change our eating habits, and we want that goal to encompass our whole family, dogs included. We can accomplish part of this goal for our dogs just by adding in Core Longevity Toppers, or CLTs, which you'll learn about in the next part, as meal toppers or treats. The great thing about our long list of CLT superfoods is that they can be added to any type of dog food (including ultra-processed diets) to improve overall well-being. You don't have to change everything at once. In fact, you may opt not to switch your dog's meals at all. Maybe for you it's the treat department that needs a major overhaul; merely replacing the expensive, poor-quality treats you're currently buying with CLTs is a great step toward taking your dog's health to the next level. You may discover, as you read, that your dog food isn't as amazing as the company or your neighbor or your vet made it sound. If you're thinking about switching dog food brands, we'll also provide the criteria you'll need to make sound brand choices based on objective nutritional parameters, not marketing hype or popularity.

Most brands of raw, freeze-dried, gently cooked, and dehydrated dog food diets are all substantially less heat-adulterated than ultra-processed pet foods and treats. From here forward, we'll refer to all of these less-processed food choices as "fresher diets" or "flash-processed diets." The type and amount of fresher dog food to feed are choices that are uniquely yours; we'll coach you through how to navigate these variables, and more, in Part III.

The conversation about dog food is due for a revolution. We hope to change how you think about feeding your dog (and yourself). There are many ways pet parents can improve their dog's nutritional status, one scoop at a time. Simple additions to your dog's bowl can create notable improvements in brain function, skin and coat health, breath, organ function, inflammation status, and microbiome balance. **Every bite of fresh, living food that replaces a bite of "fast food" (i.e., highly processed pet food) means you are taking a step in the right direction to slow aging.**

The Two Ts: Type and Timing

When it comes to "best practices" for feeding your dog, it boils down to two Ts:

Type: What type of nutrients are ideal?
Timing: When should meals be timed throughout the day?

Type: ⁵⁰/₅₀ Protein and Fat

Contrary to popular wisdom, and as we highlighted earlier, a dog's carbohydrate requirement is *zero*. That may sound absurd, given our carb-rich dietary world and the fact that most dog food is rooted in carbs. As we described in Part I, when the agricultural revolution morphed man from hunters and gatherers to farmers growing crops, dogs upregulated starch digestion to accommodate this dietary change. What happened is pretty remarkable from an evolutionary standpoint: No sooner did we start farming grains and sharing our grub with our dogs than we changed their genome. Dogs produce more amylase—the enzyme that breaks down carbohydrates—than wolves. Indeed, this shift marked a crucial step in the evolution of the wolf to dog. The natural world is sometimes cruel: evolve or die. It serves animals well to adapt to dietary and environmental changes and challenges so they can continue to live and pass on their DNA. Ancient dogs adapted to the ever-increasing carb intake of the scraps they were fed by increasing their pancreatic production of amylase. This allowed dogs to continue to reap the benefits of evolving alongside humans.

Interestingly, pet food formulator Dr. Richard Patton says even 150 years ago a dog's overall carb consumption was estimated to be less than 10 percent of their caloric intake, completely manageable for a dog's active lifestyle. In the last hundred years, since the invention of carb-rich ultra-processed dog food, a dog's involuntary carb intake has skyrocketed, which hasn't served his metabolic machinery well. Dogs can and do digest carbs, but that's not the issue. Like in humans, dogs' consumption of a diet of refined carbs over the long term will bear adverse health consequences.

Although a dog's pancreas produces enzymes that can break down carbs, that doesn't mean the bulk of a dog's calories should come from starch. It's a setup for health challenges, particularly metabolic ones that lead to systemic inflammation and obesity. Banfield Pet Hospital, which operates numerous veterinary clinics in the United States, Mexico, and the UK and is owned by Mars, Inc., reports that obesity is up 150 percent in dogs over the last decade alone.

We had a fascinating conversation with Dr. Mark Roberts about his macronutrient research on wolves and domesticated dogs. Dr. Roberts is a scientist at Massey University's Institute of Veterinary, Animal and Biomedical Sciences in New Zealand. He is most famous for his studies on how dogs would instinctually decide which foods to eat if given the choice. They don't opt for the carbohydrates. Much to the contrary, they—like wolves—*select calories coming from fat and protein first,* and the carbs a distant last. This is why many fresh food formulators say roughly 50 percent of calories (not volume of food) should come from protein and 50 percent from fat; this is the "ancestral diet" that domesticated and wild canines prefer and need.

Again, dogs are not wolves, and in diet self-selection studies, dogs do choose to consume a bit more carbs (could they have possibly developed a taste for them during the agricultural revolution?!). The range of protein, fat, and carbs these two groups of canines select is called a "biologically appropriate range of macronutrients," and this is the range we suggest you aim for, as a dog Longevity Junkie.

Now do a quick carb calculation of the dry food you're feeding: Flip over your bag of kibble, find the Guaranteed Analysis, then add up the protein + fat + fiber + moisture + ash (estimate 6 percent if you don't see a value for ash, the estimate of the mineral content of food) and subtract your result from 100. This is the amount of starch in your dog's food. Because the indigestible fiber is included in this calculation, the number you are left with is the amount of starchy carbs (that break down into sugar) in your dog food. A

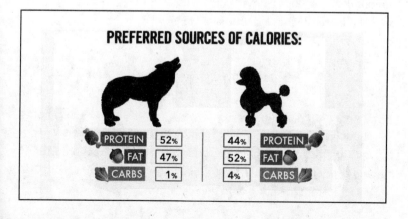

PREFERRED SOURCES OF CALORIES:

PROTEIN 52%
FAT 47%
CARBS 1%

44% PROTEIN
52% FAT
4% CARBS

biologically appropriate amount is less than 10 percent. As a veterinarian, I (Dr. Becker) have found many active dogs can tolerate up to 20 percent starch (sugar) in their diet without significant consequences; but feeding a lifetime of 30 to 60 percent starch, when none is required, will have unintended consequences.

Humans who eat mostly carbs that are made of refined flours and sugars struggle with all sorts of health issues related to inflammation. They often find relief (not to mention weight loss) when they switch to a diet with more healthy protein and fat from whole fresh foods. Contrary to popular thinking, **carbohydrates are a *nonessential* nutrient**. The glucose required by the body—especially the brain—can be synthesized from amino acids (from proteins) through a process called "gluconeogenesis." Fat yields a superfuel called "ketone bodies" that, when metabolized, actually feed the brain more efficiently than glucose, in both man and beast (humans and dogs). So it is possible for dogs and humans (and many other species) to meet their nutritional requirements without eating carbohydrates—though most of us can enjoy carbs in moderation. We should add that carbohydrates have been key to human evolution; we have side-swinging mandibles and flat molars to masticate grains, while dogs don't. There's no way we could have developed such big brains without access to carbohydrates, along with high-quality fat and protein. We aren't saying you should entirely eliminate all carbs from your diet; we are saying we must be mindful of our dogs' starch (sugar) intake if our goal is to optimize metabolic well-being and ultimately longevity.

MONEY TALKS AND BARKS

The average consumer spends $21 a month on pet food, not enough to support a quality meat and healthy fat diet. So pet food companies use minimal poor-quality meat, rendered fat, and a lot of inexpensive, feed-grade carbs to feed pets cheaply. It's important to understand that carbs are common ingredients in pet food because they are *inexpensive*, not because they are healthy. Because feeding pets correctly can be cost prohibitive, the pet food industry has convinced people that dogs are quasi-omnivores (and now can be vegans), but that comes with a price: their health—not to mention vet bills!

In addition to human studies on processed food diets, we now have studies to show what happens in the dog world: Dogs eating dry food or kibble tend to have higher inflammation and obesity rates than those eating fresh dog food. These results were duplicated in early 2020 when researchers with a fresh pet food company and the University of Florida collected body condition scores and demographic, diet, and lifestyle data on 4,446 dogs. Of these, dog owners reported that 1,480 (33 percent) of dogs were overweight or obese, with 356 (8 percent total) of dogs scored as obese. Fresh pet food in the study included commercial fresh food, commercial frozen food, and home-cooked food. According to the study's leader, LeeAnn Perry, "These types of pet food are typically characterized by the usage of whole-food ingredients that are gently cooked or minimally processed prior to being frozen or refrigerated . . . Of the 4,446 dogs in our study, 22 percent were currently being fed fresh food only, and an additional 17 percent were being fed fresh food in combination with other types of food."

The results of the study are clear: Dogs with diets consisting solely of ultra-processed pet food had a higher likelihood of being overweight or obese. And predictably, the researchers found that gradually increasing a dog's amount of exercise per week decreased the likelihood of both being overweight and obese.

We were fortunate to visit Dr. Anna Hielm-Björkman, a professor at the veterinary school in Helsinki, Finland, who studies dog metabolomics. The DogRisk program at the teaching hospital is spearheading several innovative research programs to evaluate the effects of different types of dog foods on canine health. Raw food measures better metabolically in dogs than kibble. Additionally, compared to kibble-fed dogs, canines eating a raw diet have less inflammation and lower disease markers, such as lower homocysteine levels. This rings true even for dogs that appear to be lean and healthy from the outside. Looks aren't everything. You can't see what's going on inside metabolically, physiologically, or epigenetically. Our educated guess is that millions of people and their dogs are walking around today chronically inflamed, but they don't look it. All of the experts we interviewed agreed: **Chronic, low-grade inflammation is the beginning of most diseases**.

HOW DOES INFLAMMATION MANIFEST IN OUR DOGS?

Medical conditions with an inflammatory component can be identified by their name, ending in "itis" (the suffix used for any inflammatory disease). Inflammation brings more dogs to the vet than any other reason, including these common conditions:

NAME	LOCATION	SYMPTOMS
Gingivitis	inflammation of the gums	bad breath leading to oral disease, drooling
Uveitis	inflammation of the eyes	squinting, painful eyes, rubbing
Otitis	inflammation of the ears	ear infections, redness
Esophagitis	inflammation of the esophagus	nausea, licking lips, excessive swallowing, reluctance to eat
Gastritis	inflammation of the stomach	GERD (acid reflux), vomiting, nausea, decreased appetite
Hepatitis	inflammation of the liver	vomiting, nausea, lethargy, increased thirst

Enteritis	inflammation of the intestines	nausea, vomiting, diarrhea (IBD, IBS), gas, bloating
Colitis	inflammation of the colon	diarrhea (with or without blood), constipation, anal gland issues, straining to poop
Cystitis	inflammation of the bladder	urinary tract infections, urinary crystals, straining to urinate
Dermatitis	inflammation of the skin	hot spots, sores, crusts, skin infections, itching, chewing, licking
Pancreatitis	inflammation of the pancreas	vomiting, nausea, lethargy, anorexia
Arthritis	inflammation of the joints	stiffness, joint pain, limping, decreased mobility
Tendonitis	inflammation of the tendons	knee, shoulder, elbow, wrist, and ankle pain and swelling, limping

What do all of these conditions have in common? The "itis" part of the diagnosis is fueled by pro-inflammatory foods, specifically sugar in refined carbs. The excessive starch in pet food causes ongoing elevated blood sugar levels, which in and of itself creates a pro-inflammatory state. Corn, wheat, rice, potatoes, tapioca, oats, lentils, chickpeas, barley, quinoa, "ancient grains," and other carbs found in pet foods also fuel the internal production of AGEs within the body, stimulating perpetual and progressive systemic inflammation.

When we filmed the *Dog Cancer Series,* we tracked dozens of dogs' blood glucose levels as they transitioned off of kibble and onto a raw, ketogenic diet. Their fasting blood glucose levels dropped substantially. Enough so that after receiving results, attending veterinarians sometimes called owners and expressed immediate concern about a potential hypoglycemic crisis. Much like athletes having much lower resting heart rates than their lethargic friends, dogs on a raw diet have lower fasting glucose levels compared to dogs on a starch diet.

This isn't a problem; it's a benefit. Remember, **the goal is keeping insulin and glucose low and steady in the body**, as this is the metabolically less stressful aspect of feeding real, "biologically appropriate" (low-starch) food. A dog's body secretes more insulin when glucose is greater than 110 mg/dl (6 mmol/dl). We found dogs' blood glucose often in excess of 250 mg/dl after eating a bowl of kibble. Even more concerning is how long insulin hangs around in dogs' bodies after they eat a starchy meal. A study found that insulin levels in dogs remained elevated for up to *eight hours* after eating *one* carby meal, compared to negligible insulin differences before and after a low-starch meal. Now add in carby meal number two (and possibly number three), plus starchy treats, and it's easy to see where all the "itises" originate, not to mention chronic degenerative diseases.

Is low blood sugar a risk if my dog eats a species-appropriate diet? Thankfully, only very small puppies (under five pounds) are in a high-risk category for hypoglycemia; this is the reason vets recommend feeding lots of small meals to very tiny puppies. Healthy adult dogs, even the tiny ones, have adequate glycogen and triglyceride stores to provide ongoing energy in between meals without the risk of hypoglycemia.

Domesticated dogs are still evolutionarily adapted for a state of feast and famine, as Dr. Richard Patton states; dogs produce a handful of hormones to raise blood sugar in times of famine and only one hormone, insulin, to lower blood sugar. Most modern, well-loved dogs have never skipped a meal, much less fasted more than a day or gone carb-free. In fact, most dogs have a constant stream of calories coming in all day, every day. Between multiple meals and ongoing treats, well-loved dogs have morphed into well-fed dogs whose bodies produce a constant stream of insulin, which over time taxes the pancreas and creates inflammation and metabolic stress. In Chapter 9, we will coach you through creating an optimized feeding window that minimizes metabolic stress and creates healing opportunities for the body.

The Birth and Evolution of Commercial Pet Food

The commercial dog food industry has grown exponentially since James Spratt's Patented Meat Fibrine Dog Cake. Today, most dogs eat ultra-processed commercial diets, which largely consist of by-products from the human food industry. The commercial dog food industry waged an effective "war on table scraps" when our collective perception of dogs evolved from mere pets to beloved family members, spurring the creation of a new model for feeding dogs—all-in-one commercial pet diets, purchased exclusively for them. Successful companies gained the trust of veterinarians early on, and those vets then recommended that clients feed their dogs *only* commercial pet food; human food was deemed inappropriate in light of food "specially designed" for pets. Social forces also strengthened the movement toward commercial dog food. As growing numbers of women entered the workforce in the mid-twentieth century, they had less time to prepare food for their families and dogs alike. What's more, the industrialization of agriculture produced a variety of farming innovations (like fertilizers and tractors), which allowed items like meat and grain (and any of their by-products) to be sold inexpensively and in large quantities to dog food producers.

The use of concentrated animal feeding operations (CAFOs) to raise livestock and techniques to grow cash crops such as rice, wheat, corn, sugar, and soy became increasingly common.

As farmers increased production, the price of food dropped considerably. Pet food manufacturers leveraged both extra food and food by-products from processors, the agriculture industry, and the human food supply, making commercial diets for dogs more affordable and accessible. The rapid population growth after World War II made a commercial diet for dogs a practical, convenient, and budget-friendly option as opposed to something reserved for the leisure class. Pet food companies exploited people's growing perceptions that table scraps were not safe (even the very term—*table scraps*—sounded derogatory) and implied that preparing nutritionally complete and balanced meals was complicated—and best left up to the "experts."

Although pet food companies want you to believe that their products are completely safe and, moreover, nutritional gold mines, they can be anything but. If this is the first you've heard about the pet food industry's issues, there are a few additional points you should be aware of so you can make informed brand decisions:

> ➤ There's a big safety and quality difference between "human-grade" and "feed-grade" food. Meats are inspected by the US Department of Agriculture (USDA) food inspectors and either pass (and are approved for human consumption) or fail and become feed-grade ingredients in pet and livestock feed. Realistically, "pet food" should be called "pet feed" because it's made with ingredients not approved for human consumption. Pet food companies use all feed-grade ingredients, unless the brand specifically states "human-grade" on the website. Our estimate is that less than 1 percent of canned and dry pet food is made with human-grade ingredients, which speaks volumes about the quality and levels of possible contaminants in the vast majority of pet foods. Not all feed-grade raw materials are poor choices; the problem is there's no public ranking system (such as USDA Prime, Choice, or Select) for pet feed, so it's a quality crap shoot.

> Dog and cat nutrient requirements are issued by the Association of American Feed Control Officials (AAFCO) in the United States or the European Pet Food Industry Federation (FEDIAF). Companies are supposed to follow their guidelines in order to label their products "nutritionally complete and balanced." AAFCO requires that pet food labels include a Guaranteed Analysis, a statement of nutritional adequacy, and a list of ingredients in descending order by weight. Significantly, however, the organization does not require digestibility testing or finished-product nutrient testing.

> The "best by" (expiration) date on the bag is for an *unopened* bag of food. Companies do not disclose how long the food is stable or safe to feed once the bag is opened.

> Pet food companies are not required to specify on the ingredient label any chemical preservatives or other substances added to the bulk ingredients they purchased from their suppliers.

> There are no laws or regulations requiring pet food companies to screen their products for heavy metals, pesticide or herbicide residues, or other contaminants.

> Dogs, much like humans, have different energy and nutritional requirements during different stages of life, but the vast majority of pet foods are labeled for "all life stages" (i.e., one-size-fits-all nutrition, from puppyhood to geriatric). Companies claiming "every batch is tested" in their quality-control processes should be willing to share those results with you; ask to see the test results from the batch you've purchased.

GOOD VERSUS BAD FATS

Just as there's a difference between good and bad carbs, there also are good and bad fats. Bad fats, like saturated or trans fats, stoke inflammation and often are found in highly processed fare. Healthy fats are mono- and polyunsaturated—rich in anti-inflammatory omega fatty acids. Excellent sources of healthy fats include nuts, seeds, avocados, eggs, cold-water fatty fish such as salmon and herring, and extra virgin olive oil. Fats should be unrefined, unheated, and raw. Heated fats make those dreaded ALEs (the lipid version of AGEs).

Pet food companies market carbs, including potatoes, rice, oatmeal, or quinoa, as rich sources of "energy" (aka calories). The lean, healthy protein and high quality fats that are supposed to make up your pet's diet can be negated by the calories coming from superfluous carbs. Grain-free foods often are higher in starch than grain-based foods. Their other problem is that they consist of legumes, which hold anti-nutrients like lectins and phytates. Anti-nutrients are best defined as plant-based chemicals that—while not always harmful—do not allow your cells to take in nutrients. Because they are often found in grain-free foods, it's best not to overindulge on the starchy stuff.

Another problem with grain-based foods is their potential to contain contaminant residues. As noted in Part 1, in 2020, 94 percent of pet food recalls (a whopping 1,374,405 pounds in the United States) were due to aflatoxins, a type of fungal mycotoxin definitively linked to kidney failure, liver failure, and cancer in a variety of species. Our own testing demonstrated vegan dog foods had the highest levels of glyphosate, which is passed up the food chain; leaky gut and dysbiosis are the result, leading to massive systemic inflammation (more on this phenomenon later).

A 2019 study evaluating the urine of thirty dogs and thirty cats found four to twelve times as much glyphosate than typical of human exposure, with dogs eating dry food having the highest levels. Let's not forget the 2018 study we highlighted in Part I, when Cornell University researchers analyzed glyphosate residues in eighteen commercial pet foods from eight manufacturers and found the carcinogen in every product. An ongoing study by the Health Research Institute Laboratories (HRI) of glyphosate levels in dogs and cats is also underway. Some of the results so far would make any health-conscious pet parent cringe: **Dogs have glyphosate levels thirty-two times the human average.** All of these chemically laden unnecessary carbs don't just disrupt the gut biome, they also never make dogs feel full. Fresh-feeding vets have noted many dogs on kibble appear to be ravenous, bottomless pits, never satiated and always hungry, which raises the question: Are they desperately trying to meet their biologic requirements for fat and protein by gorging on their carb-rich, sole source of calories?

Because the number of pet food companies using USDA-human edible ingredients is growing, so is the competition. Visit your brand's website: If human-grade ingredients are being used, you'll see this immediately; slogans and taglines highlighting human-grade ingredients will be plastered all over their marketing to help potential consumers understand why their food is so much more expensive. Transparency is an important differentiating factor for these companies, so often they proudly share on their website third-party, independent test results for digestibility and nutrient analysis, along with glyphosate, mycotoxin, and other contaminant test results. This is a huge step forward in building consumer trust and confidence about the quality of ingredients used in the commercial dog food you purchase.

If you don't see the information you are looking for on your brand's website, call its customer service number and ask; you'll have a hard time getting transparent companies off the phone, because they're proud of their ingredients and where the ingredients come from—they know these factors readily differentiate their products from the competition. Some fresh food companies use human-grade ingredients but don't manufacture their products in facilities approved for human food production, so their food can't be labeled as human-grade. Some fresher food companies may also choose to include some ingredients that aren't approved for human consumption (like ground fresh bone as a calcium source), which disqualifies the food from being labeled as "human grade," even though the quality is exceptional. These products are safe and healthy to feed, and these companies will happily explain this to you when you call.

Food Speaks to the Microbiome

As a reminder, most dogs can handle up to 20 percent of calories coming from starch in the diet without significant metabolic consequences, and the vast majority of dogs we meet are super-resilient. It's when we feed progressively more and more refined, high-glycemic carbs that bad things happen to all of us over time. This is about nourishing their microbiome as well. Food may be the most important element in supporting the health of your microbiome, and we cannot overstate the importance of this fact. Those microscopic bugs that inhabit the gut (and other organs, including the skin) are fundamental to health and metabolism. **The microbiome is so crucial to mammalian health that it could be considered an organ unto itself**. Our pets have their own unique microbiome that reflects their evolution and their environmental exposures, including their diets. (Fun fact: Fully 99 percent of the genetic material in your body is not your own; it belongs to your microbial comrades!) Most of these invisible creatures live within your digestive tract, and while they include fungi, parasites, and viruses, it's the bacteria that appear to hold the proverbial keys to the kingdom of your biology, as they support every conceivable feature of your health.

This incredible inner ecology helps you and your dog digest food and absorb nutrients; supports the immune system (in fact, 70 to 80 percent of both of our immune systems are located within the intestinal tract wall) and the body's detoxification pathways; produces and releases important enzymes and substances that collaborate with your biology; protects against other bacteria that cause disease; helps regulate the body's inflammatory pathways, which in turn affects risk for virtually all manner of chronic disease; helps you handle stress through its effects on your hormonal system; and even ensures you get a good night's sleep. Some of the substances these microbes manufacture are essential metabolites to your body's systems, from its metabolism to your brain function. Indeed, we essentially subcontract the synthesis of some key vitamins, fatty acids, amino acids, and neurotransmitters to these microbes.

The bacteria in your and your dog's gut produce vitamins B12, thiamine and riboflavin, and vitamin K, which is needed for blood coagulation. The good bacteria also keep things in harmony by turning off the spigots of cortisol and adrenaline—the two hormones associated with stress that can wreak havoc on the body when they are continually flowing. In the neurotransmitter realm, the gut bugs have a big hand in supplying serotonin, dopamine, norepinephrine, acetylcholine, and gamma-aminobutyrate (GABA). Just when we thought all these substances were made upstairs in the brain, we are learning otherwise, thanks to new research and technologies opening our eyes to the power of the microbiome. Although scientists are still unlocking the secrets to the microbiome and its ideal composition (and how to change it), we do know that having a diverse array of colonies is key to health. And such diversity depends on dietary choices that feed the microbes and set the stage for whether the microbiome is functioning properly. Things that can antagonize a healthy microbiome include exposure to substances that kill or otherwise negatively change the composition of the bacterial colonies (e.g., environmental chemicals, fertilizers, contaminated water, artificial sugars, antibiotics, non-steroidal anti-inflammatory drugs), emotional stress, trauma (including surgery), gastrointestinal disease, a lack of nutrients, or a biologically inappropriate diet (metabolically stressful food products).

The Human Microbiome Project, launched in 2008 to catalog the microorganisms living in our bodies, has rewritten the medical textbooks. Until the project launched, we didn't realize the immune system's command center is the microbiome itself. **The bulk of our immune system resides around our gut.** It's called the "gut-associated lymphatic tissue" (GALT) and it's significant: Upwards of 80 percent of our bodies' total immune system is attributed to the GALT. Why is our immune system largely stationed in the gut? Simple: Aside from our skin, the intestinal wall has the greatest chance of coming into contact with harmful foreign material and organisms because it is what takes in organisms from the outside world. This part of our immune system doesn't work in a vacuum. To the contrary, it communicates with other immune system cells throughout the body and sends an alert message if it encounters a potentially harmful substance in the gut. This is also why food choices are so fundamental to immune health. And all of this also relates to our dogs, too. To quote Daniella Lowenberg in her coverage of the dog's skin microbiome for the *PLOS* (Public Library of Science) blog: "A house is not a home without a dog, and a dog isn't a "D-O-double-G" without its microbial 'crew.'" Various institutions around the world are studying dog microbiome sequencing. AnimalBiome, for example, is one such company based in the San Francisco Bay Area and is dedicated to pet care through microbiome research and products. It is capable of evaluating your dog's microbiome before and after gut restorative interventions.

Italian studies have shown that feeding dogs a fresh, meat-based diet positively influences the microbiome in healthy dogs. We visited scientists Misa Sandri and Bruno Stefanon at the Department of Agricultural, Food, Environmental and Animal Sciences at the University of Udine to understand why. Clean, fresh, biologically appropriate foods nourish dogs today the same way their ancestral diet did for thousands of years; they also establish the critical foundation of cellular vitality and metabolic prowess necessary to carry dogs well past their average life expectancies. Our interview with Sandri and Stefanon provided a fascinating look into how food (and certain nutrients) can help or hinder the body's ability to rebuild and restore itself, depending upon what microbial communities are built up or destroyed in the gut. They were the first researchers to compare how a dog's microbiome changed when either a raw diet or a heat-processed diet is followed: **Raw diets create a richer, more varied microbiome.** Microbiome expert Dr. Tim Spector at King's College Medical School further highlighted the critical role of gut health, as it relates to many aspects of a dog's health span. The thing that impacted Rodney most during the interview at the London campus was Dr. Spector's concluding remarks: "Dogs and cats are given processed food their whole lives. From my recent studies, I can't think of anything worse for a microbiome than feeding high-starch, highly processed, non-diverse foods to *any* animal for sustained periods of time. This will reduce the number of microbial species in the gut, affect gene expression, and reduce the numbers of enzymes and metabolites. And that affects the immune system. The immune system is what stops allergies and cancers."

Not only will your food choices affect your dog's gut health, but we'll explain in Chapter 6 the environmental choices that impact the balance of microbes in a dog's GI tract, which impacts its immune health over time. For now, let's return to some of the marketing tactics the pet food industry employs to lure you (and your wallet).

The Truth behind the Claims

By some measures, the pet food industry is like the blue jeans industry. You can buy a pair of jeans for $30 or $300, even though it's the same denim. Some key terminology:

- ➤ "Premium" is an undefined and unregulated term (anything can be premium).
- ➤ "Vet approved" means any vet can endorse the food for any reason, including a paid endorsement.
- ➤ "Organic," "fresh," and "natural" can mean many things, as it does in the human food industry.
- ➤ Deceptive marketing is permissible with pet food products. Pictures of a perfectly roasted turkey on the bag doesn't mean the bag contains roasted turkey.
- ➤ The FDA's "Compliance Policy" allows pet food companies to use animals that have "died otherwise by slaughter" in pet food. In theory, animals that die by slaughter are healthy until they are killed. But under the Compliance Policy, animals that die from disease or other causes may be rendered and those tissues used in pet food. This is how euthanasia solution ended up in pet food that killed many animals several years ago.
- ➤ Many brands have trademarked marketing lingo to make you feel good about what you're buying—"Life Source Bits," "Vitality+," and "Proactive Health" formulas—even though we have no idea what these terms mean.
- ➤ The amount of actual supplementation in foods that promote extra benefits such as "glucosamine for hip and joint health" or "added omega-3s for skin and coat health" can be *parts per million*, literally homeopathic amounts of supplements to lure you, with no added health benefits.
- ➤ And understand what Dr. Marion Nestle calls the "salt divider": Companies know that superfoods look good on the label, but how do you know how much turmeric, parsley, or cranberries are actually in the food? Look to see where these foods are listed on the label—do they appear before or after salt? Salt (a required mineral for pets) will almost never be more than 0.5–1 percent of the ingredients, so superfoods listed after salt are for marketing show.

WHAT IS HUMANE WASHING?

Legal protections for the welfare of farmed animals are extremely limited at both the federal and state levels. The limited legal protections that do exist have not kept pace with the rapidly increasing industrialization of the meat, dairy, and egg industries. As a result, animals raised for use in these industries are subject to an array of cruel practices, which are both legal and largely hidden from public sight. Consumers, even those who care about animal welfare, are therefore unwittingly purchasing food produced through inhumane practices.

Major food producers have capitalized on this situation with marketing campaigns designed to portray meat, dairy, and egg products as more humane than they actually are—a practice known as "humane washing." Common phrases used in humane washing are "humane," "happy," "pasture-raised," "caring," "no antibiotics ever," and "naturally raised." Because statements like these are undefined by the USDA and other government entities, food producers use them loosely and often in a manner that is inconsistent with reasonable consumer interpretations. Many consumers are becoming aware of the disconnect between expectations created by humane washing marketing claims and the actual conditions of factory farms. Lawsuits alleging that food producers and pet food companies have misled consumers have been used not only to address the humane washing itself, but to shine a light on the underlying industry practices that otherwise go unnoticed.

Many people we know want to make sure that animals who become food for other animals are treated humanely, including humane slaughter. Some pet food companies have been charged with humane washing their advertising campaigns. In 2015, Rodney confronted the FDA about deceptive pictures and marketing allowed on pet food packaging during an annual AAFCO meeting in Denver, Colorado. The FDA's response was that it was freedom of speech. So our best recommendation is to never judge a book by its cover: What you see on the package *isn't* necessarily what you get. You must research your dog's pet food company like you screen schools and babysitters for your kids; dive into your research project and ask enough questions that you feel comfortable with your decision.

The definition of "processed" is endlessly debated but easily simplified with common sense. All of us—our pets included—eat processed foods (pretty much anything boxed, bagged, bottled, canned, and labeled). We recognize the convenience of kibble as reality in most people's lives, similar to us eating processed foods regularly and trying to balance things out by making better food choices when we're not so crushed for time and on the go. Human epidemiological studies determined that populations consuming the highest amounts of ultra-processed foods exhibited higher prevalence of chronic disease. This finding prompted development of food classification systems (i.e., NOVA, International Food Information Council) according to the degree of processing: minimally processed, processed, and ultra-processed.

Recently, a similar system has been proposed for pet food, with the goal of providing veterinarians with neutral terminology to discuss pet diet types with owners. The definition of ultra-processed diets for pets is the same as for humans: fractionated, recombined foods with added ingredients; in other words, dry, canned, or any other pet foods manufactured using more than one thermal or pressure processing step for the final product. According to this proposed pet food classification system, "minimally processed" commercial pet diets are fresh or frozen pet food with no or only one thermal (heat) or pressure processing step. We'll explain in Part III why we suggest slightly less-rigid definitions.

Almost 90 percent of a dog's diet is processed to some degree. As we've been detailing, it's the *ultra*-processed foods we need to be concerned about. These are the foods with ingredients that did indeed come from nature but have been mechanically, chemically, and heat processed a multitude of times and blended with other ingredients Mother Nature did not cultivate. These synthetic additives include carrageenan and other thickeners, synthetic colors, glazes and palatability enhancers, manufactured hydrogenated fats, laboratory-made vitamins and minerals, preservatives, and flavorings. Research shows it's not just the number of assaults on food that morphs plants from healthy to heinous; the duration of time and the temperature to which ingredients are heated also matter. **A typical bag of kibble is made up of foods that have been fractionated or isolated, refined, and heated an average of four times—making it, by definition, ultra-ultra-ultra-ultra-processed**.

In addition to the poor-quality ingredients and off-the-charts glycemic index in most ultra-processed pet foods, the by-products of the manufacturing process, the **Maillard reaction products (MRPs), raise *significant* long-term negative health concerns**. In recent years, commercial pet food has come under fire for containing two particularly insulting MRPs: acrylamides and heterocyclic amines (HCAs). Acrylamides are potent neurotoxins that occur when carbs (starch) undergo thermal (heat) processing. Acrylamides have raised alarms in human health circles as well; you may have heard that burned or overcooked food may raise the risk of cancer. HCAs are chemical compounds that have also been flagged as cancer-causing agents in high-heat-processed meat. These discoveries are not actually new but have been buried in the medical literature, and most certainly the ultra-processed pet food industry seeks to keep it that way. Public awareness of these damning studies has the potential to impact dramatically the estimated $113 billion in sales anticipated by 2025. Back in 2003, when scientists at the Lawrence Livermore National Laboratory in California analyzed twenty-four commercial pet foods for these carcinogenic HCAs, they found that *all but one* tested positive for the toxins. Such findings have since been repeated in study after study, including seminal investigations by Dr. Robert Turesky, a research scientist and professor of medicinal chemistry at the University of Minnesota's College of Pharmacy, where he is the Masonic Cancer Center Chair in Cancer Causation. When he detected these carcinogens in the fur of his own dogs, knowing full well he wasn't serving them grilled steaks and well-done burgers, he pointed his finger at the ultra-processed dry pet food and found the culprit.

As we previously explained, when carbs and proteins (starches and meats) are heated together (either in the body or during food manufacturing), they result in a different but equally destructive, permanent chemical reaction called "glycation": the production of advanced glycation end products (AGEs). In our interview with Siobhan Bridglalsingh, DVM, PhD, she explained the results of her 2020 study evaluating the influence of four differently processed dog food diets (canned, extruded, air dried, and raw) on plasma, serum, and urine levels of AGEs in healthy dogs. Her findings are exactly what you'd expect: Canned and extruded foods fed to healthy dogs created the highest levels of AGEs in the body, followed by air-dried diets; raw diets of course produced the least amount. Dr. Bridglalsingh noted: "Feeding dogs these diets, because of the processing methods, is very similar to a human eating a Western diet and consuming high amounts of exogenous sources of AGEs (excluding raw food)." She explained AGEs cause massive degenerative disease in dogs.

In her words: "We have found that heat processing affects the dietary AGE levels in the food and corresponds to a similar change in the free plasma total AGEs. So, we can say that high-heat-treated diets result in the formation of more AGEs in the foods, and that corresponds to an increase in the total free plasma AGEs in the circulation." When we asked Dr. Bridglalsingh if her research results changed how she thinks about pet food, her response echoes ours: "I'm much more open-minded about feeding pets homemade diets now."

What does this research mean? Dr. Bridglalsingh tells it bluntly: "It means that by feeding high-heat-processed foods to our dog, it could be the equivalent of feeding ourselves fast food all the time. And we know what the effect of us eating fast food day after day is, and we may be forcing our dogs to do this. We have a choice, but we are giving these diets to our dogs; therefore, we have a responsibility as veterinary professionals to offer them something better and something that's safer. So, if it is by feeding this diet we are predisposing them to inflammatory conditions and degenerative diseases, we can change this and do what's necessary to make a healthier food for them and possibly increase their life span and have a better quality of life by offering them a better food product."

Although this groundbreaking study is the first to evaluate dog food processing techniques and AGE formation, other pet food studies certainly have highlighted the health consequences of diet-mediated inflammation and immune system dysregulation. The test results we think every animal lover around the world should know? Our dogs are not thriving because they have *122 times higher*

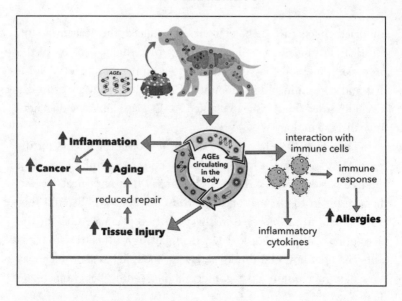

levels of these toxic compounds than fast-food-eating humans (and cats have 38 times higher levels).

We know ultra-processed diets fed to lab animals results in growth abnormalities and food allergies, but there has never been any published randomized, controlled clinical trials comparing whole-diet effects of dry, canned, and raw food on the health, disease, and life span of a population of dogs over a lifetime. The short-term studies looking at the differences between animals eating ultra-processed diets and those consuming raw, unprocessed food mirror what veterinarians are seeing clinically: raw food decreases oxidative stress, provides better nutrition because it's more digestible, positively affects the immune system by creating a more microbially diverse microbiome, and has a positive influence on a dog's DNA and epigenetic expression, including dogs suffering from skin problems. The limited number of raw-versus-processed pet food studies are already trending in the direction of what fresh feeders have reported for decades: owners feeding raw food diets report their pet has a healthier body condition, higher energy level, shinier coat, cleaner teeth, and normal bowel movements. They believe their animals have fewer health problems than animals that are fed ultra-processed food. Thankfully, feeding minimally processed or raw food helps reduce the amount of AGEs that have accumulated in tissues.

Unfortunately, the Standard American Dog Diet truly is SADD. A lifetime of highly processed food and the influx of detrimental AGEs are harmful to every tissue in the body. From musculoskeletal disease to heart disease, kidney disease, significant allergic reactions, autoimmune disease, and cancer, it's no wonder we're finally waking up to what we're feeding our pets. Not ironically, the list of problems AGEs incite in the body is the same list of reasons dogs visit the vet most often. Just as we intuitively know it's not a good idea to eat junk food all of the time, we now recognize we shouldn't feed ultra-processed food to dogs for their entire lives. Even better, maybe you've decided it's time to reduce your dog's ultra-processed food intake by 25 or 50 percent, or ditch it all together and permanently freshen up your dog's bowl. We'll help you do that.

In the dog wellness space, one type of unprocessed therapeutic dietary strategy has made big waves in the last four years. Like a human's diet, a dog's diet also can be positively manipulated to provide more calories from fat than from protein (a ketogenic diet), which is a powerful nutrition strategy for managing some forms of canine cancer. You've probably heard about ketogenic diets ("going keto"), as there's been a strong trend in people using keto diets to help manage certain medical conditions.

The ketogenic diet is similar to fasting in many metabolic and physiological respects. It severely restricts carbs and moderately restricts protein, forcing the body to turn to fat for fuel. Before it does, however, it will first burn through stored glucose and glycogen, and then the liver kicks in to produce the alternative fuel called ketone bodies. The body is "in ketosis" when ketone bodies accumulate in the blood and your resting (fasting) glucose, A1c (a measurement of glycation in the body), and insulin levels are low and steady. We all experience mild ketosis when we fast, when we get up the morning after a long, glucose-deprived sleep, or after very strenuous exercise. Ketosis has been essential to evolution because it allowed mammals to survive when food was difficult to find. It is not a metabolic state to keep your dogs in forever, but it can be used as a powerful, short-term or intermittent strategy to manage a variety of inflammatory conditions. Ketosis also has likely served an important role in dog evolution and can be leveraged today for their health, particularly in treating cancer.

When we filmed our documentary the *Dog Cancer Series*, we interviewed the folks at KetoPet Sanctuary, a nonprofit in Texas, and we met dozens of dogs fighting stage 4 cancer who were managed using a raw, ketogenic diet as a primary cancer therapy. A nutritionally balanced raw-food diet that provides about 50 percent of calories from fat and 50 percent from protein naturally creates a mild ketogenic state in most dogs. The ratio of fat to protein can be manipulated in ketogenic diets to address different metabolic needs, but KetoPet stresses the need for ketogenic diets to be fed *raw*: They have found that heat-processed fats cause pancreatitis, while raw unadulterated fats are healthfully metabolized without side effects. Pancreatitis is a real issue in small-animal medicine, and oxidized, heated fats should be avoided at all costs. **Heating fats also creates another highly toxic MRP: advanced lipoxidation end products (ALEs), which many toxicologists believe are the most damaging of all to organ systems.**

Dr. Mark Roberts's research on dogs self-selecting these same approximate macronutrient ratios (50 percent of calories coming from healthy fat and 50 percent from protein) confirms that dogs maintain some ancient, innate metabolic wisdom—given a choice, domesticated dogs prefer to consume fat and protein as their primary energy sources, which results in lowered metabolic stress and enhanced physical and immunologic performance in animal models. It makes sense. We simply need to implement meals with macronutrient ratios in the ballpark of Mother Nature's beautifully honed bandwidth, developed over the last ten thousand years.

Timing: Honor the Biological Clock

Health through food alone is one thing. But health through food *and meal timing* is an exponentially more effective means to achieve optimal health and longevity. As we heard repeatedly in our search for the best wisdom about mammalian vitality, the healthiest food in the world, eaten at the wrong time of day, becomes a physiologic stressor. That's right: **"How much you eat and what you eat are very important, but *when* you eat might be even more important."** Those are the words from Satchidananda Panda, PhD, at the Salk Institute. He's blazing the trail toward better health through better meal timing. Calories can't tell time, but your metabolism, cells, and genes certainly can. By honoring our pooch's innate metabolic machinery, we're doing our part in reducing dietary stress; and by respecting a dog's ancient eating window, we can reap the profound health benefits of a balanced circadian rhythm, an innate internal clock that has regulated our sleep/wake cycles for thousands of years.

We all have a biological clock, whether we're a man, a woman, or a dog. It's technically called a "circadian rhythm," and it's defined by the pattern of repeated activity associated with the environmental cycles of day and night. These are rhythms that repeat roughly every twenty-four hours, and they include our sleep-wake cycle, the ebb and flow of hormones, and the rise and fall of body temperature that correlate roughly with the twenty-four-hour solar day. A healthy rhythm's command of normal hormonal secretion patterns is especially important, from hormones associated with hunger cues to those that relate to stress and cellular recovery. When your rhythm is not synchronized properly, you don't feel 100 percent, and you'll probably be grumpy, tired, hungry, and prone to infection because your immune system isn't fully operational. Those who have flown across time zones and experienced jet lag know all too well the uncomfortable feeling of disrupting the circadian rhythm.

As we'll see in Chapter 6, your sleep habits determine your circadian rhythm. Accordingly, sleep deprivation can do a serious number on our appetite. For example, the major hormones in control of our appetite, leptin and ghrelin, determine when we do and don't eat, and they operate constantly. Ghrelin tells us we're hungry, and leptin lets us know when we're full. The emerging scientific research on these digestive hormones is breathtaking: The findings show that not getting enough sleep disrupts the natural order of these hormones, which then impacts hunger and appetite. One study found that when people sleep only four hours a night two nights in a row, their hunger increased by 24 percent. They also craved high-calorie sweets, salty foods, and starch. It can be deduced that—when one is tired—the body looks for fast energy, and that can be found in high-processed, refined carbs.

Don't for a minute think a dog's circadian rhythm is not as big of a deal. It is. We paid a visit to the Salk Institute in Southern California, where we met Professor Satchidananda Panda at his Regulatory Biology Laboratory (Panda Lab) to discuss the effects of timing on food consumption. An animal's innate circadian rhythm dictates when food is nourishing and healing, or metabolically stressful; and caloric restriction (or "intermittent eating or fasting") can add years to a pet's life. Panda's research demonstrates that by picking up the food bowl and limiting treats to match an animal's biological clock, many of the most common age-related metabolic diseases can be avoided.

When we tell pet owners that it's okay if their healthy dog doesn't want to eat for a day or skips a meal, they are surprised. But dogs do not need to eat two or three square meals a day with treats in between (and neither do humans). Like us, dogs are equipped to fast. They should occasionally fast for part of the day, in fact, to hit their metabolic reset button.

Intermittent fasting, sometimes referred to as time-restricted feeding (TRF) for animals, has a long history dating back thousands of years (there's a reason why most religions incorporate fasting into their practice). Hippocrates, the Greek physician who lived in the fifth and fourth centuries BCE, and who gave us the Hippocratic Oath, was one of the fathers of Western medicine. Hippocrates was a big proponent of fasting for health. Among his writings he proposed that both disease and epilepsy could be treated with complete abstinence from food and drink. The Greco-Roman philosopher Plutarch, in a work titled "Advice about Keeping Well," said, "Instead of using medicine, rather fast [for] a day." Avicenna, a great Arab physician, often prescribed fasting for three weeks or more. The ancient Greeks used fasting and calorie-restricted diets to treat epilepsy, and this practice gained a revival in the early twentieth century. Fasting also has been used to detoxify the body and purify the mind to reach complete natural health. Even Benjamin Franklin gave us his opinion that "the best of all medicines is resting and fasting."

Fasting comes in many forms, but its fundamental effect on the body is the same. Fasting activates the hormone glucagon, which counterbalances insulin to keep your blood glucose levels balanced. Here's a visual to really grasp this concept. Picture a lever or seesaw: When one person goes up, the other goes down. This analogy is frequently used to simplify or explain the biological insulin-glucagon relationship. In your body, if the insulin level goes up, the glucagon level goes down, and vice versa. When you give your body food, your insulin level rises, and your glucagon level decreases. But the opposite happens when you don't eat: Your insulin level goes down, and your glucagon level rises. When your glucagon level rises, it triggers many biological events to take place—one of which happens to be autophagy, that cellular cleaning mechanism we discussed earlier. This is why temporarily denying your body and your dog's body nutrients through the safe practice of time-restricted eating (or feeding, when it comes to dogs) is one of the best ways to boost the integrity of your cells. Giving your dog all their daily calories during a set time is better for their physiology. In addition to preserving "cellular youth" and slowing the aging process, this way of eating also increases energy and fat burning and reduces risk factors for conditions like diabetes and heart disease. These benefits occur because fasting triggers autophagy, otherwise known as cellular house cleaning.

Mark Mattson, a neuroscience professor at Johns Hopkins School of Medicine and former chief of the Laboratory of Neurosciences at the National Institute on Aging, is a prolific researcher in this area. He has collaborated with Dr. Panda on the research and published extensively in medical literature. He is particularly interested in how fasting can improve cognitive function and reduce the risk of developing neurodegenerative diseases. Dr. Mattson has conducted studies in which he subjected animals to alternative-day fasting, with a 10 to 25 percent calorie-restricted diet on the in-between days. According to him, *"If you repeat that when animals are young, they live 30 percent longer."* Read that sentence again. By changing when animals eat, we can extend their life span—by a lot! **It's not just more time; it's more time with better health and less disease.** Dr. Mattson even found the animals' nerve cells were more resistant to degeneration when following this protocol. And when he performed similar studies in women over the course of several weeks, he found that they lost more body fat, retained more lean muscle mass, and had improved glucose regulation.

Ironically, one of the mechanisms that triggers these biological reactions is not just autophagy but *stress*. During the fasting period, cells are under a mild stress (a healthy, "good" kind of stress) and they respond to that stress by enhancing their ability to cope with it and, perhaps, to resist disease. Other studies have confirmed these findings. Fasting done correctly reduces blood pressure, improves insulin sensitivity, boosts kidney function, enhances brain function, regenerates the immune system, and increases resistance to diseases in mammals across the board.

Fasting is natural to a dog's physiology as well, and they benefit in the same fashion. Some dogs will naturally fast themselves, which can alarm their owners. But their self-imposed fasting behavior mimics what would happen in nature, and it gives the digestive system a break and allows the body to rest, repair, and restore itself. A growing number of animal experts recommend fasting healthy dogs (that weigh more than ten pounds) one day a week, perhaps just giving them a recreational bone to gnaw on that day. For some, the thought of their dogs skipping a meal doesn't sit well, and research verifies many owners consider their pets as family members and believe that food restriction may lead to possible distress in the animal. As a consequence, these owners may not comply with TRF suggestions or weight-loss programs because they don't want to stop giving snacks or restrict the quantity of food. But creating healthy routines is a necessary part of creating a healthy life for our dogs. Calorie tough love (which we prefer to call "healthy food boundaries") is part of creating healthy eating habits for many dogs. In Part III, we'll give a very long list of ultra-low-calorie snacks and treats you can feel great about feeding throughout your dog's set "eating window." Just to be clear, fasting means withholding food, never water.

WHO'S TRAINING WHOM?

"But you don't know my dog!" We hear this a lot. Dogs are very perceptive creatures of habit. Most guardians don't realize they unknowingly create all sorts of annoying, food-related behaviors in their dogs, from the dancing and barking that may start when you open the fridge, to the constant whining when you sit down to eat, to the vomiting of bile if their dinner isn't exactly on time (more on this in Chapter 9). It's important to recognize that you have (perhaps unwittingly) fostered your dog's responses, behaviorally and physiologically.

That's right: You are responsible for the furry food monster's behavior in your house. If you have inadvertently created a treat beast, you can mindfully reshape these behaviors, starting today. It will take time and patience, but we believe the only way to improve unwanted behaviors is to consistently and appropriately address them through thoughtful, positive behavior modification. Dogs naturally repeat behaviors that serve them, meaning if there's a chance they can get what they want by behaving a certain way, they'll do it again (and again, sometimes becoming more obnoxious to make sure you're paying adequate attention and responding the way they want). If there's no response from you (literally none: no verbal acknowledgment, no eye contact, literally zero reaction), in a short time your dog will stop the behaviors that are no longer effective at eliciting the response from you that she wants; her human-training skills have now met their match! We know many dogs who woke up their owners at night with obnoxious "hangry" behavior, until their owners corrected these patterns. The Purina Institute found Beagles fed twice a day increased their nighttime activity by approximately 50 percent compared to dogs who eat once daily, so TRE can even improve the amount of restorative sleep you both enjoy.

In Part III, we'll give you some ideas for introducing a TRE window into your dog's life to enhance her rest-repair-rebuild cycle, a key piece of building a Forever Dog. **The combination of feeding your dog the right nutrients in the right amount at the right time is the magical trifecta of biological wins.** We hope you're now convinced that giving Fido free license to eat whatever, whenever, with carby treats in between, is a biological bomb. Treats are important, but treat portion and timing are even more important.

Fresh Is Best, Which Means It's "People Food from the Fridge"

One of the commercial dog food industry's greatest accomplishments is convincing a generation of dog owners that feeding "people food" to dogs is both nutritionally and socially unacceptable. But this sentiment has met its match in twenty-first-century science. Not all people food is bad for dogs. In fact, human food is the best-quality food your dog will ever eat: It's passed inspection! The quality of people food is much better than the feed-grade fare most dogs eat. But the *type* of human food you provide to your pooch is critically important. And of course, **you don't feed human food directly from the table; instead you use biologically appropriate people food for dogs to create balanced meals, or as training treats or toppers in their food bowl**.

In short, we encourage you to feed real, healthy, fresh foods to *everyone* in your family (just remember, no onions, grapes, or raisins for dogs). A balanced approach to feeding our dogs a blend of fresh-from-your-fridge and commercially made fare is a great choice, and we'll map out exactly what that means in Part III. We want you to choose the food, feeding style, and pet food companies that resonate with your personal food philosophy and your wallet.

LONGEVITY JUNKIE TAKEAWAYS

➤ The pet food industry is poised for radical change as pet owners demand healthier, fresher alternatives to traditional kibble and canned dog food.

➤ Most brands of raw, freeze-dried, gently cooked, and dehydrated dog food diets are all substantially less heat-adulterated, than modified, heat-processed kibble and canned pet food.

➤ In the last hundred years, since the invention of carb-
  rich, ultra-processed "dog food," a dog's involuntary carb
  intake has skyrocketed, to the detriment of his metabolic
  machinery.

➤ The ratio: Roughly 50 percent of your dog's calories should
  come from protein and 50 percent from fat; this is the
  ancestral diet that domesticated and wild canines prefer
  and that best serves the twin goals of health and longevity.
  Feeding 30 to 60 percent starch over a lifetime, when none
  is required, has unintended health consequences.

➤ Studies now show that the more kibble dogs eat, the greater
  her likelihood to be overweight or obese and to show signs
  of systemic inflammation ("-itis" diagnoses).

➤ Food is not just information for your own cells, tissues,
  and systems; it's a key piece of information for your gut's
  microbiome, and it factors mightily into the strength and
  function of our metabolism and immune system.

➤ Timing matters: *When* your dog eats is as important as
  what and *how much* he eats. Calories can't tell time, but
  one's metabolism, cells, and genes certainly can. When the
  body's circadian rhythm is synched properly, the diet is more
  nourishing and less metabolically stressful. Intermittent
  fasting—what we call time-restricted feeding (TRF)—can
  be a powerful tool for dogs just as it is for humans. Healthy
  dogs do not need to eat three square meals a day with treats
  given liberally in between. When they don't want to eat a
  meal or treat, don't fret. Not only is it normal (assuming they
  aren't sick) but it's beneficial. The combination of feeding
  your dog the right nutrients in the right amount at the right
  time is the magical trifecta of biological wins.

5

The Triple Threat

How Stress, Isolation, and Lack of Physical Activity Affect Us All

> In times of joy, all of us wished we
> possessed a tail we could wag.
>
> —W. H. Auden

Tina Krumdick tells a powerful story of her dog Mauzer's journey to health:

"Mauzer had chronic diarrhea. For a year I brought her to a local vet who ran almost every test under the sun and found nothing. It always felt like guesswork. Every time I left the office, I had a different type of kibble and more medications to try and firm up her poop. Nothing worked. I felt like they were treating her symptoms and not finding the cause. After the last appointment I remember driving home with a $60 bag of kangaroo kibble and wondering, 'What if it's the food causing the problem?'

"A friend of a friend recommended seeing Dr. Becker. I wasn't sure because she was over an hour's drive away and I heard she was not a 'typical' vet. Then Mauzer started pooping undigested kibble and blood; it was coming out of her like a firehose. She lost seven pounds in a week, and I thought she might not make it. I made an appointment, because perhaps an atypical approach was just what she needed. Dr. Becker walked in, sat on the floor, and Mauzer crawled in her lap. I knew we were in the right place. Nothing they had tried worked. Dr. Becker diagnosed her with malabsorption, after a blood test.

"Dr. Becker talked to me about feeding a raw diet. Real food made sense. She also went on supplements to help her recover from her malabsorption. Within days her poop was normal and she started gaining weight. No more blood. She was no longer lethargic. Then I started rotating between buffalo, venison, and turkey proteins she hadn't had before. The variety didn't upset her system, it healed it. I used frozen blueberries for treats. I would also set aside a little of what I was making myself for dinner to add to her dinner. All fresh and healthy. Her coat changed from dull to shiny. She was more energetic. Her eyes were bright. She thrived. Thousands . . .

I had spent thousands chasing a problem that started with what I was feeding her. Ironically, food caused the problem, and food also cured her."

Mauzer's positive transformation reflects what many other dogs experience with attention to diet. There's no doubt that Mauzer's diet was stressing her body. Reducing that metabolic stress helped turn things around. Although stress comes in many different forms, it has a single result when it's relentless: ill health. And as stressed out as that makes you, we have to tackle this subject.

An Epidemic of Stress

If we were to ask a room full of people to raise their hands if they occasionally—perhaps persistently—experience anxiety, inner agitation, fatigue, fear, irritability, and feelings of overwhelm, our guess is that a lot of hands would go up. If we could do the same for their dogs, asking them to bark about their stress levels, there would be a lot of barking.

We all can agree that we humans are especially stressed out today. More than 30 million Americans take antidepressants, and since the 1990s this percentage has increased by more than 400 percent in the United States. Suicide rates have increased in nearly every state since the turn of the twenty-first century. One-fourth of American adults say they experience insomnia, which may lead them to use sleep aids like pills. Although we like to think social media brings us together, it can have the opposite effect: More than three in five Americans admit to feeling lonely, and only half of Americans say they enjoy regular, fulfilling socialization with other people.

We're also not moving much anymore, which compounds our physical and mental stress load. A mere 8 percent of adolescents in the United States meet the suggested sixty minutes of daily exercise, and less than five percent of adults get the recommended thirty minutes. And thirty minutes is just a minimum recommendation. Most Americans spend more than half the day seated or lying down. We are nowhere near the ancestral average, either: modern hunter-gather tribes like the Hadza of Tanzania live by foraging, so their women typically walk around three miles a day, while men walk around seven miles daily.

For most of the past, our survival depended on the exercise and movement necessary to complete daily tasks. Hunter-gathers had to forage and hunt for food, and those actions required walking and running—often with animal companions. Our brains became bigger and stronger because of exercise, and we began to create communities, rules and laws, trade and the sharing of resources, and other social intricacies that ensured our safety and survival. Dogs were often an essential part of this.

Sitting may just be killing many of us. How so? A 2015 meta-analysis and systematic review published by the *Annals of Internal Medicine* demonstrated that a sedentary lifestyle is correlated to premature deaths from all causes. Not being sedentary may also prevent illness and early death. Witness a 2015 study, which found that getting out of the seated position once an hour for two minutes of light movement was linked to a 33 percent decrease in a premature death by any cause. Other lengthy studies show that movement may decrease the risk for colon, breast, lung, endometrial, and meningioma (a type of brain tumor) cancers, likely due to the fact that exercise helps minimize inflammation. The less chronic inflammation, the lower the risk that cells will become cancerous.

This is equally true for our dogs: if they moved more like their ancestral counterparts or if they could choose how much time they would like to spend running, sniffing, and moving outside, they'd reap the benefits of lowering their risk for premature aging and disease, including depression. Although we don't often talk about depression in dogs, they are suffering alongside our skyrocketing rates with depression and its less-serious cousin, anxiety.

Dogs may not experience depression the same way we do, but anyone who has witnessed a dog go through a traumatic experience—the death of an owner or family member, a natural disaster, exposure to loud noises, a geographic move, or a change in family dynamics (e.g., new spouse, new baby, divorce)—knows that dogs are fully capable of exhibiting sadness, lethargy, or other atypical behaviors in response to these stressors. They may refuse walks, stop eating, start barking, act withdrawn, or otherwise behave in ways we may deem "unbecoming." Vets often prescribe anti-anxiety medications in these circumstances, and they are the same medications we take—Paxil, Prozac, and Zoloft. We believe there's a better way than a prescription pad.

Anxiety and aggression are common problems in dogs. According to the *Journal of Veterinary Behavior*, up to 70 percent of behavioral problems in dogs are attributable to some form of anxiety. Although abuse and neglect certainly contribute to anxiety and behavioral problems in dogs as they do in humans, other sources of canine stress can be more subtle and insidious: confusing, aversive training techniques; prolonged alone time; poor sleep; and lack of exercise. These causes are highly treatable without drugs.

It's really no surprise that, in the past decade, exercise has been acknowledged as an effective strategy both for treating anxiety and depression and for preventing these conditions altogether. (How long will it take for that study to make the canine rounds and reach the average vet?) A 2017 study followed forty thousand adults who hadn't been diagnosed with any mental health concerns for at least eleven years. The study found that regular leisure-time exercise significantly reduces the risk of depression, which happens to be the leading cause of disability worldwide. This suggested to researchers that as little as one hour of physical movement a week can prevent 12 percent of future depression cases. Then came a 2019 Harvard study that raised eyebrows. This one, involving hundreds of thousands of people (signs of a good study), suggested that jogging for fifteen minutes daily (or walking or gardening for slightly longer) could help prevent depression. Using the progressive research measure called Mendelian randomization, scientists could identify cause and effect relationships between changeable risks factors such as the amount of exercise and its impact on health concerns like depression. The findings are truly remarkable, suggesting that "enhancing physical activity may be an effective prevention strategy for depression."

Powerful therapy indeed. But most of us are not getting the message, and we've roped our dogs into the same, sedentary ruts. **Dogs need exercise each day based on their individual personality, body, and age (so we can't provide generic guidelines that would apply to all dogs). Daily movement therapy, as we like to call it, helps dogs feels calmer, reduces restlessness, improves sleep, and can enhance how dogs interact with one another.** There are direct mental and physical benefits, of course, but exercise has direct effects on stress itself. Animal behaviorists have long recommended exercise for common behavior problems in dogs, as it's one of the most effective tools we have. But it's also the most profound tool we have for managing stress. The anti-anxiety benefits for humans last for four to six hours after twenty minutes of aerobic exercise. If repeated daily, there's a cumulative affect. The same is true for animals.

Lab rats voluntarily choose to exercise on a running wheel, if one is provided. Research shows their endorphins (which reduce pain and increase feeling of well-being) last for many hours, only returning to typical levels ninety-six hours after the exercise session. The effects of exercise on the brain last much longer than the time exercising, and for a dog with hyperactivity disorder or anxiety, even short exercise sessions can be a godsend for improving long-term quality of life for both of you. Daily, rigorous exercise rewires the fight-or-flight stress response so prevalent in our canine companions. Exercise also changes the chemistry of the brain, including altering and promoting brain cell growth to create a calmer state. Our dogs are suffering as much as we are. Exercise has protective benefits against the detrimental effects of large amounts of stress in animals, including fear and anxiety. This is probably why *New York Times* writer Aaron E. Carroll says "Exercise is the closest thing to a wonder drug." The burning question: Are we giving our dogs the opportunity to move their bodies as often and as much as they need?

We also must mention **the importance of choice**. Yes, dogs deserve and should be given the power of making independent decisions. We spent a whole podcast talking about this subject with Dr. Alexandra Horowitz, senior research fellow in the Department of Psychology at Barnard College, Columbia University. The bestselling author of *Inside of a Dog: What Dogs See, Smell, and Know*, Dr. Horowitz specializes in dog cognition and, like us, is a big proponent of letting dogs enjoy being dogs. How often do we let our dogs choose which way to go on a walk, left or right? We aren't talking about teaching dogs to walk on a leash or being obedient; we're talking about delegating certain decisions to our dogs—including our dogs in decisions we'd normally make for them. **Creating a partnership, not a dictatorship.** When our dogs let us know there's an overwhelming need to sniff something in a direction we didn't intend on going, how often do we honor their wishes? How long do we let them sniff before yanking the leash? Affording dogs a measure of control over what happens to them, where they want to go, and what they want to do is more important than we think. The truth is, many dogs never have a say in how their life goes. Giving dogs more choice in all realms of life is a gift; in giving them agency, we respect their need to participate actively in their own well-being (and ours!), which in turn improves their confidence, quality of life, and, ultimately, appreciation and trust in us.

"Nose work" (also called "scent work") is an enriching and mental exercise you do with your dog. Scent exercises are exceptionally beneficial for reactive dogs or traumatized dogs who tend to melt down or shut down on walks. All dogs can benefit from a variety of "brain games" and mental stimulation that a nose "sniffari" provides while outside or on walks. These activities cater to their innate desire to sniff and help to alleviate stress.

Premature Aging in Stressed Dogs

Premature aging is something we'd all like to avoid. The global antiaging cosmetics market was valued at $38 *billion* in 2018 and by most accounts will reach $60 billion by 2026. Life in general inflicts normal wear and tear on all of us; but toss in serious anxiety, toxic stress, and perhaps some depression and that age clock will tick faster and faster. Consider how quickly presidents age and turn gray after four or eight years in office. People who have dealt with major stress, anxiety, or serious depression often emerge looking uncharacteristically older, like they've weathered a storm, and it's written all over their faces. Stress can indeed do a number on our physical appearance—but it does a double number on our insides. And the same holds true for our dogs.

There's plenty of research on premature graying, one of the outward physical hallmarks of aging, in humans. Among the biggest factors associated with premature graying are oxidative stress (biological rusting), disease, chronic stress, and genetics (a predisposition to graying thanks to our genes). The combination of underlying genetic forces and a stressful lifestyle lowers resistance to stress in the hair follicles and melanocytes, the cells that create your hair color.

We now have similar research on dogs. A 2016 study published in *Applied Animal Behaviour Science* reported a significant correlation between anxiety and impulsivity, and premature muzzle graying in young dogs. Although it's common for the fur along a dog's muzzle line to turn gray with age, this phenomenon isn't common in young dogs (less than four years old). In this particular study, the authors reviewed case studies from an animal behavior practice and noted that many dogs with premature graying also demonstrated anxiety and impulsivity issues. Similarly, an earlier study reported an association between certain behavioral measures, such as hiding or running away, and elevated cortisol levels in the dog's fur.

Cortisol, you'll recall, is the body's hormone classically tied to stress levels—higher cortisol indicates higher stress levels (and, in turn, inflammation). Cortisol does serve a beneficial purpose, as it directs and buffers the immune system and primes the body for attack. Cortisol works beautifully for short-lived and easily resolved threats. The attack of our modern-day lifestyles, however, is unrelenting, and cortisol is pumping around the clock. Continual exposure to excess cortisol over time can lead to increased abdominal fat, bone loss, a suppressed immune system, and a heightened risk for insulin resistance, diabetes, heart disease, and mood disorders. In our dogs, those mood disorders are often described as behavioral problems such as aggressiveness, destructiveness, fearfulness, and hyperactivity.

The cortisol levels documented in these dogs reflected chronic emotional reactivity. These two studies are not outliers; plenty of others have identified potential symptoms of canine anxiety and impulsivity. An anxious dog, for instance, may whine or prefer to stay close to her owner. A dog with impulse issues may struggle to focus, bark endlessly, or exhibit signs of hyperactivity. The authors of the 2016 study suggested that **when a dog is assessed for anxiety, impulsivity, or fear issues, muzzle grayness should be considered**. Too much gray at an early age could indicate the dog is under too much stress—a condition that can be reversed. Which raises a good question: What *is* stress?

The Science of Stress

In physics, the term "stress" means the interaction between a force and the resistance to counter that force, but we all know that the word means a lot more than that today. We throw the word "stress" around every day: *I'm so stressed out!* being the most popular. The symptoms of stress are universal and reflect a wide spectrum, from feeling moody and irritable to experiencing a racing heart, upset stomach, headache, or full-blown panic attack. Some people feel a sense of impending doom. As a reminder, stress is a necessary (and unavoidable) component of life. It helps us avoid danger, focus, and respond instinctively to those famous fight-or-flight situations. When we're on edge, we're more alert and responsive to our environment, which can be very beneficial. But prolonged stress can have long-term consequences, both physically and mentally.

GOOD STRESS, BAD STRESS

Just as there are good and bad fats and carbs, the same is true with stress. Examples of good stress include dietary habits like fasting, which stresses cells ever so slightly to produce effects in the body that ultimately are positive. Exercise also stresses the body in a good, health-promoting way. But there are plenty of examples of bad stress that can lead to unwanted outcomes. Research shows, for example, that yelling at dogs or physically punishing them causes the chronic secretion of stress hormones in dogs, which has been linked to shortened life spans.

We owe the term "stress," as it is used today, to an Austro-Hungarian-Canadian endocrinologist from the early part of the twentieth century. In 1936, János Hugo Bruno "Hans" Selye defined stress as "the non-specific response of the body to any demand made upon it." Dr. Selye proposed that when subjected to persistent stress, both humans and animals can develop certain life-threatening afflictions (such as heart attack or stroke) that previously were thought to be caused by the culmination of certain physiological variables. Now recognized as a founding father of stress research (famous enough to be emblazoned on a Canadian stamp), Dr. Selye highlighted the impact that everyday life and experiences can have not only on our emotional well-being but also on our physical health.

You may be surprised to learn that the actual word "stress," as it relates to emotions, was not part of our mainstream vocabulary and commonly used language until the 1950s and the onset of the Cold War. That's when we replaced the label "fear" with the label "stressed out." Plenty of research since Dr. Selye's days confirms over and over again that persistent stress is a real insult to our physiology. We can even measure stress's effects on physiological systems as chemical imbalances in the activities of the nervous, hormonal, and immune systems. It can be measured by disturbances in the body's day-night cycle—your circadian rhythm. Scientists also have measured changes in the brain's physical structure resulting from stress.

The tricky thing about stress is that our physical reactions don't change all that much, regardless of the type or magnitude of a perceived threat. Whether it's a truly life-threatening stressor, or just a long to-do list or an argument with a family member, the body's stress response is essentially the same. First, the brain sends a message to the adrenal glands, which promptly release adrenaline, also referred to as epinephrine. The adrenaline increases your heart rate and the blood flow to your muscles, preparing your body for flight. When the threat is gone, your body returns to normal. But if the threat remains and your stress response intensifies, then another series of events is triggered along the hypothalamic-pituitary-adrenal (HPA) axis. This pathway entails multiple stress hormones with the hypothalamus directing a lot of traffic. The hypothalamus is a small but important governing region of the brain that plays a vital role in controlling many bodily functions, including the release of hormones from the pituitary gland housed inside. It's the part of our brain that links the nervous and endocrine systems, regulating many of the autonomic functions of our body, particularly metabolism. Well known as the seat of our emotions, it's also the headquarters of emotional processing. The moment you feel stressed out (or whatever you want to call it: nervous, worried, tense, anxious, overwhelmed, etc.), the hypothalamus sends out a stress coordinator—corticotropin-releasing hormone (CRH)—to start a cascade of reactions that culminates with cortisol peaking in the bloodstream. While we have long understood this biological process, newer research reveals that the mere *perception* of stress can trigger inflammatory signaling from the body to travel to the brain, priming it for hyper-response. This process unfolds similarly in our dogs. It's a mechanism that has been conserved throughout the animal kingdom for millennia. A 2020 study by a group of Finnish researchers looked at pet anxiety in nearly fourteen thousand dogs and concluded, "Some of these behavior problems have been suggested to be analogous, or possibly even homologous to human anxiety disorders, and the study of these spontaneous behavior problems arising in a shared environment with people may reveal important biological factors underlying many psychiatric conditions. For example, canine compulsive disorder resembles human OCD on both phenotypic and neurochemical level." Put another way, **when you're stressed, your dog is probably stressed, too**.

TRAINING METHODS INFLUENCE LONG-TERM STRESS

Dog training is an unlicensed and unregulated profession, with no minimum educational requirements, standards of care, or consumer protection. "Buyer beware" doesn't come close to summarizing the sometimes irreparable damage inflicted on dogs through abusive training methods. If you want to modify some of your dog's behaviors, be very aware that the trainer you choose and the training method he/she uses will impact your dog's health and could trigger (or quell) chronic anxiety, fear, and aggressive behaviors. **The researchers we interviewed unanimously agreed aversive training methods compromise the long-term welfare of dogs.** There are safer, kinder, and wiser approaches than yelling, hitting, choking, and shocking. For your dog's mental well-being, stick with trainers who subscribe to science-based training methodologies (if you need guidance getting started, see our list on page 407 in the appendix).

The next question becomes: How do we gain better control of our stress—for ourselves and for our furry family members? The answer will surprise you, but in addition to adequate high-quality sleep and exercise, both of which play mightily into our ability to manage stress through their multiple effects on our biology, it takes guts.

Sleep and Exercise on Stress

Anyone who has gone long periods without getting a good night's sleep or without forcing the heart to pump faster in active physical movement knows the outcome: moodiness, irritability, and basically feeling like crap. Like humans, dogs need adequate sleep and exercise. But a dog's sleep and exercise habits aren't exactly like ours. For starters, dogs don't sleep for one single stretch at night and then stay awake all day, like we do. They tend to doze off whenever they want, often out of boredom, and can wake up quickly to jump into action. They will nap throughout the day in addition to nighttime sleep for a total of between twelve and fourteen hours per day (with slight variations based on age, breed, and size); they spend only about 10 percent of their snoozing time in the rapid eye movement (REM) phase, when their eyes roll under closed lids as they react to dreams. Because their sleep patterns are more irregular and lighter (less REM), they require more total sleep to help compensate for their lack of deep sleep (we spend up to 25 percent of sleep in REM).

But similarities do exist between dogs and humans in the importance of sleep. In studies on canine sleep, dogs experience short bursts of electrical activity, called "sleep spindles," during non-REM sleep, like we do. The frequency of these sleep spindles also has been linked to how well dogs retain new information they learned immediately prior to their naps, mirroring studies in humans, where quality sleep is linked to how well we remember newly banked information. **Sleep spindles are how we, and our dogs, consolidate our memories**; when sleep spindles happen, the brain is shielded from outside, distracting information. Dogs with more frequent sleep spindles during a snooze session are shown to be better learners than dogs with less frequent sleep spindles. These results track findings in humans and rodents.

Regardless of differing sleep patterns between dogs and humans, however, sleep is equally essential to both for refreshing the brain and body and keeping the biology running smoothly and metabolically intact. And just as lack of restorative sleep can lead to health challenges, so too can excessive sleep, which can signal conditions such as canine depression, diabetes, hypothyroidism, and possible loss of hearing.

Exercise has long been proven to be essential to well-being— whether you're a sapien, a canine, or any other mammal. It's arguably the most powerful, scientifically proven way to support healthy metabolism (e.g., controlling blood sugar and overall hormonal balance, as well as keeping inflammation in check). But it also helps maintain muscle and ligament tone and bone health, boost blood and lymph circulation and oxygen supply to cells and tissues, regulate mood, lower perceived stress levels, increase heart and brain health, and contribute to more restful, sound sleep. Indeed, sleep and exercise go hand in hand. We know we're not the first to tell you that these two key habits are vital to good health, but we sometimes forget just how necessary they are for our canine friends—albeit in different doses, forms, and intensities.

It Takes Guts to Be Cool, Calm, and Connected

Earlier we described the importance of the microbiome's health—the community of microbes dominated by bacteria that live within and on us (and our dogs). Initial work on the microbiome's contribution to wellness primarily focused on digestive health and immune stability. But now the science, especially as it relates to canines, is exploring the connection between the microbes in your dog's digestive tract and their brain—a complex relationship supported by evidence. Indeed, the communication between the gut and the brain may affect mood, which then influences behavior.

Intestinal microbes participate in myriad different functions, from synthesizing nutrients and vitamins to helping us digest our food and preventing us from metabolic dysfunction, including obesity. The good bacteria also keep things in harmony by turning off the cortisol and adrenaline spigots—the two hormones associated with stress that can wreak havoc on the body when they are continually flowing. We don't think of our gut and brains as being strongly connected (like our fingers and hands), but the "gut-brain axis" is incredibly relevant for this conversation. The brain and the gut communicate through the chemicals produced by gut bacteria, which then travel through nerves and hormones.

We've all experienced the connection through nerve-racking experiences that leave us feeling sick to our stomach or, worse, dashing to the bathroom. The vagus nerve is the primary conveyor between the hundreds of millions of nerve cells in our central nervous system and our intestinal nervous system. That's right: Our nervous system is composed of more than just our physical brain and spinal cord. In addition to the central nervous system, we all have an intestinal (enteric) nervous system that is built into the gastrointestinal tract. Both the central and enteric nervous systems are created from the same tissue during fetal development, and both are connected via the vagus ("wandering") nerve, which extends from the brain stem all the way down to the abdomen. It forms part of the involuntary (autonomic) nervous system and directs many bodily processes that don't require conscious thinking, such as maintaining heart rate, breathing, and managing digestion. The sympathetic nervous system is our body's fight-or-flight system—the one that quickens our pulse and blood pressure to shunt blood to the brain and muscles, away from digestion. It keeps us alert and mentally adept. The parasympathetic nervous system, on the other hand, is our rest-and-digest system that allows us to rebuild, repair, and sleep.

The potential stress-causing effects of microbes in the gut—or the absence thereof—were first explored in the study of so-called germ-free mice. These are mice that have been specially raised without normal gut inoculation, thereby allowing scientists to study the effects of missing microbes or, conversely, expose them to certain strains and then document changes in behavior. A landmark 2004 study revealed some of the first clues to the bidirectional interaction between the brain and gut bacteria. It demonstrated that germ-free mice respond to stress in dramatic fashion, evidenced by altered brain chemistry and elevated stress hormones. This condition could then be reversed by giving them a strain of the bacterium *Bifidobacterium infantis*. Since then, multiple animal studies have explored the relationship between the influence of gut bacteria on the brain and, specifically, emotions and behavior. The effects of these chemicals and hormones produced in the gut depend on which bacteria are present, because different bacteria make different chemicals. Certain bacteria manufacture chemicals that have a calming effect, while others may promote depression and anxiety. For example, in many studies, feeding mice certain probiotic bacteria (*Lactobacillus* and *Bifidobacteria*) decreased anxiety and depression because it facilitated the production of chemicals sent to the part of the brain that manages emotions. In short, certain gut bacteria impacted mood and behavior.

All of this biology we just described happens in dogs, too. In fact, most of the research to sort out how gut bacteria talk to the brain was first studied in nonhuman animals (especially mice). But get this: It's now been documented that the dog gut microbiome and its relationship to the canine brain more closely resembles our gut microbiome in composition and the functional overlap. How a dog's gut-brain axis works is similar to ours. This new science helps explain how these tiny organisms influence the emotions of dogs, possibly causing anxiety that may lead to aggressive or other undesirable behaviors.

One study that provides a striking illustration is from Oregon State University. In 2019, researchers sampled gut bacteria from thirty-one dogs confiscated from a home where they were made to fight in a dog-fighting ring. Dogs were divided into two groups (aggressive and not aggressive toward other dogs) after being observed by researchers. By carefully sampling the dogs' poop, they analyzed their gut microbiome and noticed higher levels of aggression-causing bacteria in dogs displaying the behavior. They also found that particular gut bacteria could be linked to aggression and anxious behaviors. And root causes of the anxiety can be traced to the gut and its microbial inhabitants.

The idea that the characteristics of the gut's microbial composition can reflect anxiety levels and behavior is gaining lots of traction in research circles, as scientists map out which species correlate with which outcomes. In turn, scientists are learning which diets support which microbial profiles. If a dog's diet affects the kinds of bacteria that live inside their gut, then which diet supports a healthy gut and its beneficial downstream effects? A few studies comparing meat-based raw-food diets to kibble have concluded that dogs fed raw diets have a more balanced growth of bacterial communities and an increase in fusobacteria (a good thing). In one study, dogs who were fed a raw diet for at least one year were shown to have a richer and more even microbiome compared with kibble-fed controls. We also know that the microbiomes of fresh-fed dogs pave the way for increased secretion of the "happy hormone" serotonin (healthier gut-brain axis) and less cognitive decline, with better control of *Actinobacteria*, which is associated with cognitive decline and Alzheimer's disease in both dogs and humans.

It's empowering to know we can attempt to address an array of issues—anxiety, stress, depression, gut inflammation, and even cognitive decline—through the gut. Constantly in flux, the microbiome may be affected by diet, medications (i.e., antibiotics and nonsteroidal anti-inflammatory drugs, or NSAIDs), and the environment. Research shows it can take months to healthfully restore your dog's microbial community after one round of antibiotics. Even a week of daily NSAIDs (Deramaxx, Previcox, Rimadyl, Metacam, etc.) can significantly affect gut health. We certainly aren't suggesting you discontinue your dog's pain medication, but instituting a damage control plan to mitigate GI consequences from the long-term use of many pharmaceuticals is something many prescribing veterinarians are now implementing. Rebuilding and optimizing your dog's microbiome is best achieved through simple lifestyle habits that play into a body's overall physiology and that define our—and our dog's—average day: sleep, movement, diet, and exposures. Age also factors into the equation because gastrointestinal microbial diversity, which is key to gut health, is impacted by age. The older we and our dogs grow, the harder it is to maintain that diversity. You'll see in Part III that we encourage you to add tiny amounts of a wide variety of healing foods to your dog's bowl because the science is clear: **the more microbially rich and diverse your dog's gut flora is, the healthier they will become**.

Fecal Microbiome Transplants Tell Secrets

Microbiome Restorative Therapy, or MRT, is a fancy name for a fecal transplant: taking a sample of the microbiome of a physically and emotionally healthy donor via the stool and giving it to an ill patient after it's been filtered. While this may sound alarming, MRT is an ancient practice, with roots reaching back to Africa, where mothers have long used this practice to save their babies dying of cholera. Fast-forward several centuries: The top US hospitals now use this rudimentary treatment to save patients with life-threatening *Clostridium difficile* (*C. diff*) infections. As a veterinarian, I was aware of fecal transplants being used to treat humans with life-threatening GI infections, but I had not considered using it in veterinary practice until I met Felix.

Felix was a ten-week-old yellow Labrador puppy that contracted parvovirus despite being vaccinated for the disease. His owners had spent over $10,000 trying to save his life, but despite their efforts, Felix was hospitalized at a specialty center ICU and was fading quickly. A few days later, his owners were notified that Felix was unable to stand and they should consider humane euthanasia. That's when Felix's mom, Whitney, called me. She asked if I had "any lifesaving tricks up my sleeve" that she might try before their scheduled euthanasia that afternoon. I told her about fecal transplantation and suggested she bring to the hospital fresh feces from one of her other incredibly healthy, raw-fed Labradors. If the attending doctor would allow, they would make a slurry and administer it to Felix as an enema, to flood his infected GI tract with millions of his sister's beneficial microorganisms.

It worked. Felix stood up a few hours after the transplant. It was the beginning of his recovery. At that moment, everyone involved in Felix's care recognized the power of poop. Since then, research has gone on to make some amazing discoveries: Transplanting feces from healthy mice to depressed mice cures their depression. Fecal transplants inoculated from thin mice to obese mice result in weight loss, and transplanting feces from amicable dogs to aggressive dogs improves behavior. We're just beginning to understand the breadth of health issues that can be treated effectively using this simple, ancient, proven health procedure.

Speaking of Poop: "Coprophagia" is the medical term for eating poop. Most dogs do it on occasion throughout their lives, if given the opportunity. This gross habit can provide clues about the health and needs of our dog's microbiome. Dogs will innately try to fix their maladies with the tools and resources available to them in their environment, including consuming free poop. Dogs seek out and eat different kinds of poop for different reasons. Researchers believe some dogs are seeking a source of probiotics to fix a digestive issue. Dogs may eat feces if there is partially digested food present in the stool, or if they are lacking a nutrient or substance that's present in the feces they've found (for instance, rabbit poop is a naturally occurring rich source of supplemental digestive enzymes). Sometimes coprophagia is behavioral: Dogs will eat their own feces under certain circumstances. If your dog participates in this unsettling habit, try rotating through a variety of probiotic and digestive enzyme supplements until you find a beneficial combination. If your dog regularly eats wild animal poop, it's important you bring your dog's stool sample to the vet once a year to check for internal parasites that can be passed up the food chain.

Integrity of the Gut Lining Is Key

The health, strength, and function of the gut lining are important—it separates a body's insides from the outsides and its potential dangers. Whether dog or human, the gastrointestinal tract is lined with one single layer of epithelial cells, from the esophagus to the anus. Every mucous-producing part of the body, including the eyes, nose, throat, and gastrointestinal tract, can easily attract and let in pathogens. That means the body has to protect these entry points.

The intestinal lining, the largest mucosal surface, has three main jobs. First, it serves as the avenue through which a body obtains nutrients from foods. Second, it blocks potentially harmful particles from entering the bloodstream, including chemicals, bacteria, and other organisms or pieces of organisms that can pose a threat to health. Finally, the intestinal lining plays a direct role in the immune system through classes of proteins called "immunoglobulins" that bind to bacteria and foreign proteins and prevent them from attaching to the gut's lining. These proteins are antibodies released from immune system cells on the other side of the gut lining, and they are transported into the gut via the intestinal wall. This function ultimately allows the body to usher pathogenic (bad) organisms and proteins through the digestive system and be excreted in the feces.

The inability to absorb nutrients from the gut is one of the major causes of permeability issues or a so-called leaky gut, where substances that shouldn't be allowed to cross over into the body can gain illegal entry and provoke the immune system. These junctions determine, to a large extent, the overall level of systemic inflammation. **It's well documented now that when an intestinal barrier is compromised, it can lead to a spectrum of health challenges, a myriad of symptoms, and ultimately a lifetime of chronic disease.**

This lining also shares important relationships with the intestinal flora—and your diet. Processed foods can release bacterial toxins that normally hang out in the gut as part of the microbiome; when the gut's wall is compromised by a leaky gut, those toxins can escape into the bloodstream and wreak havoc in circulation. Meanwhile, the composition of a healthy gut flora can become adverse and imbalanced, resulting in a condition we defined earlier called dysbiosis. Many things can disrupt your dog's GI flora: antibiotics, veterinary pesticides (flea and tick medications), steroids, and other veterinary drugs (NSAIDs). Some of these insults are temporary and necessary, but the biggest culprit is what's hidden in your dog's ultra-processed food: Glyphosate residues, mycotoxins, and the constant stream of AGEs our dogs consume all contribute to dysbiosis and a disrupted microbiome. One animal model study demonstrated that, **in addition to creating leaky gut, Maillard reaction products (MRPs) increase the amount of potentially harmful bacteria in the gut**. It's no wonder so many animals eating processed diets have gut issues, not to mention immune system dysfunction. Remember, much of a dog's immune system is in the lining of her intestinal tract, which is now perpetually compromised. When we hear from pet owners about their dog's GI issues, food and environmental allergies, behavioral or neurologic problems, and autoimmune diseases, we often suspect dysbiosis and leaky gut as the root cause. But the fix is simple: a more natural diet that not only nourishes the microbiome's composition and functionality but also maintains the gut's integrity.

Dysbiosis occurs silently, with no outward symptoms until there's a systemic immunologic reaction. That's when the itching, scratching, and GI symptoms become apparent. In both dogs and humans, dysbiosis is associated with obesity, metabolic diseases, cancer, and neurological dysfunctions, to name a few. Unfortunately, the most prescribed antibiotic for GI problems in dogs, metronidazole (Flagyl), profoundly exacerbates dysbiosis. Metronidazole kills fusobacteria—the bacteria dogs need to digest protein—and opportunistic pathogens are allowed to take hold. In addition to creating irritable bowel symptoms (including a drop in fusobacteria), there is an increase in segmented filamentous bacteria, SFB. This can trigger the epigenetic expression of interleukin-6 (IL-6) and other inflammatory pathways that create systemic inflammation in the body. It can also trigger the epigenetic upregulation of the Th17 gene, which in turn leads to atopic dermatitis and other inflammatory skin conditions. Research also indicates that fuso-bacteria are decreased in kibble-fed dogs. The DogRisk group discovered similar patterns in gene expression between healthy and atopic Staffordshire Bull Terriers that ate raw or dry food. Raw food appeared to activate the expression of genes that had anti-inflammatory effects.

Microbiome projects are well underway around the world, as this promising area of research is still in its infancy. What has become clear is that the gut microbiome helps direct many physical and psychological processes in our pets. This is why we can't stress enough that the key to having a Forever Dog lies in a healthy gut. We'll return to this conversation within the context of diet. The food we feed our dogs, and its role in the gut, may be one of the most overlooked when we talk about behavior. Just as feeding kids too many refined, processed foods high in sugar and additives can turn them into over-aroused, hyperactive, cranky children, the same is true for our pups.

Soil Support

Dogs have innate wisdom and wise instincts, finely honed over years of evolution, that guide them in making healing choices—when given the opportunity. But most dogs aren't given regular opportunities to make the instinctual choices that would heal their own bodies.

"Zoopharmacognosy," the word to describe animals' self-medicating behavior, derives from ancient Greek: "zoo" (animal), "pharmaco" (remedy), and "gnosy" (knowing). Animals knowing what they need, and when.

Zoopharmacognosy has been well described in wildlife literature for decades. Dr. Michael Huffman brought more attention to this fascinating area of research in the 1980s, when he first published

his extraordinary observations of wild chimpanzees carefully se-
lecting medicinal plant parts to address different maladies (his
TEDx Talk on the subject is fascinating).

We asked Dr. Huffman about domestic dogs and "pica" (the medical term for animals eating nonfood items, such as soil, clay, and
toilet paper). He smiled and explained that **domesticated animals
still have ancient instincts that serve them very well, but
most aren't allowed to naturally express behaviors that help
balance their bodies.** We make most of our dogs' decisions for
them, and we don't give them adequate opportunity to sniff, dig,
and discern what organic substances they need to correct their biome imbalances or a trace mineral deficiency. Dogs are often left
with very few in-home options—licking carpet fibers, chewing bits
of tissue they can steal from the trash can, and eating the occasional weed they can harvest between sidewalk cracks. That's a far
cry from nature's botanical medicine cabinet they co-evolved with.
Even worse, they're often punished for expressing these desperate
cravings—a recipe for anxiety, indeed. We're not suggesting you
turn your dogs loose in the country all afternoon to lick limestone
to correct a calcium deficiency, but we do recommend evaluating
your dog's behavior to better understand what they're searching
out, and why. Don't let your dog investigate unsafe areas, including chemically treated environments. But *do* let your dog be a dog:
give them time and space to nibble grasses, lick dirt, paw to discover the root they're searching for, sample a specific weed, and
taste the clover hidden in and among lawn sod. If your dog is fever-
ishly looking to ingest organic substances (followed by vomiting),
your dog has microbiome issues or is sick. Otherwise, the grass
she's going for, in between the sidewalk cracks, may be her only
opportunity to self-select what she's yearning for. Let her enjoy it.
You can learn more about using applied zoopharmacognosy with
your dog at www.carolineingraham.com.

For these reasons, our close friend Steve Brown launched the Canine Healthy Soil Project. Steve is well aware that the puppies he's

raised with early access to healthy, chemical- and toxin-free soil are *significantly* healthier than litters raised by other breeders exclusively indoors, with strict hygiene protocols that create near-sterile living environments during the first eight weeks of life. Healthy soil is highly biodiverse; one gram may contain 10 billion microbes and two thousand to fifty thousand-plus species that directly speak to a dog's microbiome. Research backs the requirement for puppies to have access to microbially rich soil very early on in life for long-term immune well-being. Because most puppies have not been given this opportunity, Steve wants to provide it now, in the form of soil-based microbiome support that helps nourish the microbes modern dogs are missing. His goal is to help dogs develop balanced, biodiverse microbiomes by providing food and beneficial bioactive metabolites from a wide variety of soil-based microbes. These microbes assist in the development of a diverse microbial population in a dog's mouth, gut, and, especially when applied topically, on the dog's skin and coat. With two years of research and development under his belt, he's blown away by the overwhelmingly beneficial results, especially pertaining to behavior, allergies, obesity, diabetes, oral health, breath, skin and coat, arthritis, and brain function.

Down to Earth

In recent years, there's been talk about the power of so-called nature therapy, which is about going out into nature for fresh air and more peace of mind. The movement stems from the Japanese tradition of *shinrin-yoku*, or forest bathing: immersing yourself in the sights, sounds, and smells of nature. *Shinrin-yoku* was developed in Japan during the 1980s and has been promoted by its Department of Forestry since 1982 as a public health initiative. Reported research findings associated with this healing practice include therapeutic effects on immune function, cardiovascular health, respiratory disease, depression, anxiety, and hyperactivity disorders. Researchers attribute some of the beneficial immune effects of being in nature to inhaling molecules called phytoncides secreted by trees and plants to protect themselves from pests and diseases.

Humans need direct contact with the earth, and studies show that not having it—something that our sedentary, interior modern lifestyles promote—can harm our health. Making a conscious effort to spend time outside in nature can promote positive physiological changes, as well as increase a sense of general well-being.

Have you ever wondered how animals know there's an earthquake coming before it happens? The answer is Schumann resonances: the electromagnetic vibration of the Earth. Yep, it's true. The Earth has an energy force to which dogs (and you, science shows) are sensitive. Dr. Abdullah Alabdulgader's team published a fascinating paper in *Nature* evaluating the Earth's magnetic forces and how they influence mammalian autonomic nervous system (ANS) responses. Their research strongly confirms that daily ANS activity in mammals responds to changes in geomagnetic and solar activity. This explains how energetic environmental factors can influence psychophysiology and behaviors in different ways (think full moons and earthquakes!). Animals are especially sensitive to Schumann resonances, which, at 7.8 hertz, are nearly identical to alpha brain wave frequencies (those linked to calmness, creativity, alertness, and learning).

Studies conducted at the Halberg Chronobiology Center at the University of Minnesota (Dr. Franz Halberg coined the term "circadian") demonstrate that many human and animal wellness measures are directly related to the Earth's rhythms and resonance. When biorhythms are disrupted, confusion and agitation were some of the first symptoms. **We believe—and we preach—that your dog needs to roll around on the ground, dig holes, sleep under a tree, or do anything that directly connects them with the earth.** Preferably several times a day.

In the 1960s, 90 percent of human doctor visits were related to acute injury, infectious disease, or childbirth. That's all changed:

95 percent of medical appointments today are due to lifestyle or stress-related complaints or conditions. Clearly, our modern lifestyle is not promoting good health. This is true with pets as well. Fifty years ago, most vet visits occurred because of sudden illnesses or injuries that required medical assistance. Today, the pet patients we see need treatment for conditions includeing GI problems, allergies and skin issues, musculoskeletal issues, and organ dysfunction. It's an epidemic. The best way to ground yourself and your dog is to get outside and touch earth: Go for a walk. Every animal, if given the chance, uses Earth's magnetic fields to assist them in their day-to-day lives; dogs can even find their way home when they're lost. In fact, studies suggest that animals sense that certain parts of their bodies need to make contact with the earth to benefit them physiologically, so they roll around or lie on their bellies or touch their nose to the ground. The problem is we don't always give them this opportunity. All of the Forever Dogs we've met spent a lot of time outside every day. The more time you can spend outside in a safe environment with your dog, allowing them to sniff, dig, roll, run, move, and play, the more grounded (and, dare we say, happily fulfilled) they will be.

LONGEVITY JUNKIE TAKEAWAYS

➤ An epidemic of toxic stress—the kind that puts too much pressure on the body and leads to unhealthy outcomes—is plaguing both the human and canine worlds, yet there are simple strategies to combat stress that don't have to come from a prescription bottle.

➤ Exercise is an antidote for stress, anxiety, depression, and feelings of loneliness—whether you're a dog or human. Just two minutes of movement an hour has been associated with a reduced risk of death from any cause. Daily movement therapy, as we like to call it, helps dogs feels calmer, reduces restlessness, improves sleep, and can enhance how dogs interact with one another.

➤ Is your dog prematurely gray? And behaving badly? These can signal too much stress. And when you're stressed out, chances are your dog is, too. Your dog's physical and emotional cues are symptoms to alert you to an underlying issue that needs attention.

➤ Our intestinal biomes, including those of our dogs, factor strongly into our health and can be influenced for good or bad by dietary choices, level of physical movement, sleep quality, and environmental exposures. This also means that we can have a positive impact on the gut's microbiome through lifestyle choices.

➤ Dogs love to make some of their own decisions for their health, which can include things like sniffing and foraging on (unsprayed) grasses. Such behavior can have a positive impact on their microbiomes. Too many dogs, however, are not given enough opportunities to explore nature, take sniffaris, dig in the dirt, and literally ground themselves.

6

Environmental Impact

The Difference between a Muddy
Dog and a Dirty Dog

> The dog is a gentleman; I hope to
> go to his heaven, not man's.
>
> —Mark Twain

In 2010, I (Dr. Becker) had a feline patient with asthma who required the progressive use of an inhaler to try to manage her uncontrolled asthma symptoms. When I dug deeper into why the cat's asthma became unmanageable over several months, I learned the owner had signed up to become a representative of a popular direct-to-consumer home-scenting business. She had become a top sales rep by hosting lots of home-scenting product parties where all of the highly scented wax-melting pots, plug-in diffusers, and other irresistibly scented home sprays were displayed and available for purchase. Every room of her home had some kind of scenting device. Simultaneously, her cat's asthma became severe—enough that she was hospitalized. When the cat's mom removed all the volatile organic compound (VOC)-spewing contraptions from her home, her cat's asthma quieted down. Her dog's chronic conjunctivitis, eye discharge, and paw licking also resolved.

Indeed, environment matters. A lot.

The Hazards of Modernity That Lessen Longevity

When we were kids growing up, seat belts were optional (especially
if you rode in the way-way back), people could smoke wherever
they wanted (airplanes included), the drinking age was eighteen,
trans-fat-filled margarine was favored over butter, and we'd microwave foods in their plastic containers (remember TV dinners?).
We'd also ride bicycles and ski without helmets, drink water from
the phthalate- and lead-leaching hose in the backyard (remember
that metallic taste?), and sunbathe without sunscreen (baby oil was
preferable). Today these behaviors are banned below a certain age,
totally prohibited, or at least heavily discouraged. We did a lot of
other things as kids that would be frowned upon today or otherwise seen as unhealthy. Every generation identifies new dangers
to avoid or regulate, and we fully expect to see additional scrutiny
and testing in the future, particularly with regard to chemicals and
their associated products. But sadly, regulation lags far behind such
investigations and probably always will.

By the time we find out about the potential hazards of a substance (or a behavior or activity), many of us already have experienced exposures and their effects. The Environmental Protection
Agency (EPA), European Union (EU), and World Health Organization (WHO) have each promised to accelerate their efforts to gather
data on "contaminants of emerging concern," and the Centers for
Disease Control and Prevention (CDC) has established a national
system for tracking environmental hazards and the ailments or illnesses they may cause. The National Institute of Environmental
Health Sciences (NIEHS), established by the NIH in 1966, also
undertakes and supports research, but it is not involved in the bio-
monitoring. It's unlikely that any regulations will be implemented
fast enough to ensure our safety and that of our dogs.

In many respects, we live in a safer world today compared to several generations ago. Fewer people are injured in car crashes, wars,

and natural disasters; the global disease burden has fallen thanks to better medicine, including public health and sanitation measures. We're more likely to die at an old age than from an injury or a sudden heart attack at forty-two. But in the realm of exposures, our modern lifestyle still leaves much to manage and bring into safer territory. **We cannot push our longevity limits any further if we don't control and mitigate our exposure to pollution in all its forms, including what we inhale, absorb, or even take in through our eyes (like blue light at night), and what we hear that disrupts our sense of serenity.**

When you think of pollution, your mind probably goes to belching smokestacks from factories, smoggy cityscapes, bottles of solution with the skull-and-crossbones warning labels, car exhausts, landfills, and plastic-filled oceans. We tend not to think about the more insidious, invisible pollution that we and our pets encounter every single day. Take a moment to consider all the comforts around you that reflect modernity. Go back to this morning and think through your day—from the cosmetics, toiletries, personal products, and cleaning supplies you used to the furniture you sat on; the electronics you engaged with; the lawns, carpets, and hardwood floors you crossed; the indoor air you breathed; the water you drank; the mattress on which you slept; the clothes you wear; the fragrances you smelled; and the excess noise and light you experienced. The list goes on—and we haven't brought food into the picture yet. For purposes of this chapter, we're going to stay focused on the toxic exposures we encounter aside from food.

In fact, here's a great way to gain a sense of your everyday non-food exposures. Check any of the following boxes if your answer is yes:

❑ Do you drink unfiltered tap water (and fill your dog's bowl with the same water)?

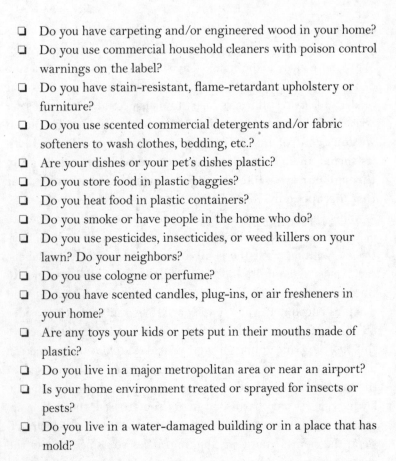

- ❏ Do you have carpeting and/or engineered wood in your home?
- ❏ Do you use commercial household cleaners with poison control warnings on the label?
- ❏ Do you have stain-resistant, flame-retardant upholstery or furniture?
- ❏ Do you use scented commercial detergents and/or fabric softeners to wash clothes, bedding, etc.?
- ❏ Are your dishes or your pet's dishes plastic?
- ❏ Do you store food in plastic baggies?
- ❏ Do you heat food in plastic containers?
- ❏ Do you smoke or have people in the home who do?
- ❏ Do you use pesticides, insecticides, or weed killers on your lawn? Do your neighbors?
- ❏ Do you use cologne or perfume?
- ❏ Do you have scented candles, plug-ins, or air fresheners in your home?
- ❏ Are any toys your kids or pets put in their mouths made of plastic?
- ❏ Do you live in a major metropolitan area or near an airport?
- ❏ Is your home environment treated or sprayed for insects or pests?
- ❏ Do you live in a water-damaged building or in a place that has mold?

The higher your score, the greater the potential toxic burden you may carry. Now think of a day in the life of a dog. Imagine putting a video camera on his or her head to record what kinds of exposures occur. They'd be similar to your own, because you and your dog share living experiences, such as drinking the same water, enjoying the same couch, inhaling the same air, and brushing up against your professionally washed clothes and scented skin. Dogs are even more exposed to these pollutants by virtue of their proximity to the ground, their lack of protective clothing, and their less frequent bathing to rinse off chemicals and contaminants. You're five or six feet up, but your dog is inches from the ground, and often sleeps where chemicals loom in flooring and where all the invisible, airborne particulates eventually land. The fumes from household cleaners fill the air, and certain products routinely off-gas chemicals from their inherent materials (the smell of a new vinyl shower curtain, for example). Your dog's nose is up to 100 *million* times more sensitive than yours. Household dust, which hides on floors and in corners, typically is filled with potential toxins that accumulate like tumbleweeds. Older homes may have lead paint that's inhalable, ingestible, or lickable on windowsills or near floorboards where paint crumbles or chips.

Outside, dogs love the soft plush of the grass, but if it's been treated, those wet paws and noses are taking in a good dose of carcinogens. It all makes for quite a heavy load, or body burden. Studies dating back twenty years now show that household pesticides—insect repellents; various products to control ants, flies, cockroaches, spiders, termites, and plant/tree insects; herbicides, including those used by professional lawn services; weed-control products; and flea-control products (including indoor foggers, flea collars, flea soaps or shampoos, sprays, dusts, or powders)—all correlate with a stunning increase in risk for certain cancers in both children and pets. In one of the most alarming early studies done by a large group of researchers from around the world, led by Dr. Elizabeth R. Bertone-Johnson at the University of Massachusetts, **your pet's risk of developing canine malignant lymphoma may go up by about 70 percent if they're exposed to lawn pesticides (specifically pesticides used by professional lawn care companies).** Fully 1.8 million people tuned in to watch the educational video about lawn chemical risks for dogs that we posted on Facebook many months ago; the response from surprised, newly informed pet parents was overwhelming. People were shocked, and many sprang into action, rethinking their lawn-care practices.

Another similar study from Purdue University found a strong link between chemically treated lawns and increased risk for canine cancers. The Purdue study looked specifically at the risk of bladder cancer in Scottish Terriers, which have a history of developing bladder cancer far more frequently than other breeds. Their genetic predisposition toward developing this cancer makes them ideal as "sentinel animals" for researchers, because they require far less exposure to a carcinogen than other breeds before contracting the disease. Lo and behold, the Purdue group found that the greater the exposure, the greater the risk: The occurrence of bladder cancer was between four and seven times *higher* in the group exposed to the chemicals. The similarity between dog and human genomes could lead researchers to find the gene in humans that renders them susceptible to developing bladder cancer.

This study was particularly noteworthy because it highlighted an important truth about chemical mixtures we use in lawns and gardens: So-called inert ingredients may be to blame. Billions of pounds of untested chemicals reach our lawns and gardens every year, and while it's easy to point to known villains like DDT or glyphosate, it's much harder to pinpoint other hidden culprits right under our noses and feet—literally.

The Body Burden

As we briefly noted in Part I, those of us living in industrialized nations now have hundreds of synthetic chemicals in our bodies, accumulated from food, water, and air. They make up what's called the "body burden," and they are stored in almost every tissue, including fat; cardiac and skeletal muscle; bones, tendons, joints, and ligaments; visceral organs; and the brain. How these chemicals are stored depends on their chemical nature; fat-soluble toxins such as mercury get stored in fatty tissues, whereas water-soluble toxins like perchlorates (which can land in water supplies) are typically excreted through urine after they've gone through the body. Many toxins are fat-soluble, which means the more fat you have, the more toxins you retain. Another bad piece of news is that toxins can trigger water- and fat-retention. When the body is overloaded with toxins, inflammation naturally ensues, and our bodies respond by retaining water in a bid to dilute both fat-soluble and water-soluble toxins.

To reiterate: the vast majority of these chemicals, many of which come from plastics, have never been adequately tested for health effects. Chemicals from plastics are absorbed by the human body— 93 percent of Americans six years of age or older test positive for bisphenol A (BPA), which you know now is a chemical derived from plastics and proven to adversely affect our biology—especially the hormonal (endocrine) system. Some of the other compounds found in plastic have also been found to affect hormones or have other detrimental health effects.

In the United States, mass spectrometry is a method used to screen for more than 170 environmental pollutants, including organophosphate pesticides, phthalates, benzene, xylene, vinyl chloride, pyrethroid insecticides, acrylamide, perchlorate, diphenyl phosphate, ethylene oxide, acrylonitrile, and many others. Test results using eighteen different metabolites from a urine sample can be helpful to determine what your personal body burden is—how many chemicals your body harbors and which kind. We are working to make these tests available for pets, too. But even without these specialized tests, scientists have plenty of evidence to show the burden we all harbor, from unborn fetuses to old dogs. Contamination from toxic chemicals is widespread.

Some of the same researchers in the Purdue study also have documented chemicals in the urine of dogs living in households with both treated lawns *and untreated* lawns, making a case that even people who don't douse their lawns still could be exposing their pets (and themselves) to adverse chemicals through grassy travels (e.g., neighborhood walks, parks) or simply the drift that occurs from neighboring lawns. Vapors from herbicides travel a lot farther than you think—up to two miles, though most vapors drift within an eighth of a mile. But that's far enough to reach a few houses down the road.

At Texas Tech's Institute of Environmental and Human Health, scientists have documented unlikely sources of BPA and phthalate exposure in dogs: toys and training devices like bumpers that dogs love to chew. These items are made with plastics that leach chemicals, which are also categorized as endocrine disrupting chemicals (EDCs) for their ability to damage the hormonal system. These same chemicals are known to adversely affect humans as well and are associated with early puberty in girls as a result of hormonal impacts.

> **Tip:** The two most offensive lawn treatment chemicals that are bona fide carcinogens are 2,4-Dichlorophenoxyacetic acid (2,4-D) and glyphosate (the key ingredient in Roundup). Check your lawn-treatment chemicals; and when in doubt, take it out. There are safer alternatives to almost every toxic product or service out there.

GIVE YOUR DOG A FOOT BATH

Your dog's paws are like damp little Swiffer pads, picking up all kinds of allergens, chemicals, and other pollutants. Keep in mind that dogs sweat only from their noses and the pads of their feet. So those damp little pads can collect a really heavy load of irritants. Often you can dramatically reduce the amount of time your dog spends licking and chewing her feet with a fast, easy foot soak. Depending on the size of your dog, you can use the kitchen or bathroom sink or a bathtub.

Fill the sink or tub with a few inches of water, enough to cover those paws. Our favorite solution is povidone iodine (available at drugstores and online), an organic, non-irritating solution that's safe, nontoxic, antifungal, and antibacterial. Dilute the povidone iodine with water to the color of iced tea (medium brown). If the solution is too light, just add a bit more of the iodine. If it's a bit dark, add more water. Simply let your dog stand in the solution for two to five minutes. You don't have to do anything to their paws; the solution will do the work for you. If your dog is nervous about being in water and talking or singing doesn't calm him down enough for the soak, offer treats. Then pat dry and go! Repeat every two to three days.

Gassy grasses are one hazard on the home front, but so is what we encounter indoors, where we spend more than 90 percent of our time. Unless you're rolling around and burying your face in a heavily treated lawn, indoor environments can be more toxic than the outdoors in many ways. Numerous studies over the past decade, including one meta-analysis published in 2016 by a consortium of US institutions that made mainstream news, have definitely shown that household air can be a toxic cocktail—often filled with dust that contains chemicals toxic to the immune, respiratory, and reproductive systems. The cocktail also contains VOCs such as formaldehyde and combustion by-products such as soot and carbon monoxide. It's clear that VOCs—including the ones that give cars a "new car smell"— are major reasons why the indoors are so toxic. VOCs easily transform into gas and can combine with other chemicals, forming compounds that can irritate, inflame, or more when they're inhaled or absorbed through the skin. They are found in a wide variety of products: cologne, carpet adhesives, glues, resins, paints, varnishes, paint strippers and other solvents, wood preservatives, foam insulation, bonding agents, aerosol sprays, cleansers, degreasers and disinfectants, moth repellents, air fresheners, room-scenting plug-ins, stored fuels, hobby supplies, dry-cleaned clothing, and cosmetics.

Even though you may be taking every precaution to minimize these airborne chemicals, choosing eco-friendly alternatives wherever possible, our advice is to buy a high quality air purifier. Airborne toxins are major causes of pollution in the home, and you may not even notice them. Different air purifiers are equipped to handle different pollutants, from smog, smoke, and particles, to chemicals, gases, and fumes, to mold, viruses, and bacteria. Others are able to assist with everything. More than 90 percent of particulates are small enough to be effectively removed by a HEPA (high efficiency particle absorption) filter. Air purifiers are able to alleviate symptoms of allergies or asthma. Other ways to mitigate those symptoms are to change air-conditioning filters often and clean the ducts annually. However, the fastest and easiest way to keep toxin levels low in your home is to ventilate your house regularly. Open the windows!

> **Tip:** Don't panic at the thought of replacing all your household products today to live a cleaner lifestyle with your dog. There are simple things you can do to reduce exposures. Switch to fragrance-free, environmentally friendly laundry soap next time you run out. That couch and dog bed with chemically treated upholstery? Toss an organic blanket or covering made with natural fibers over it. Your questionable carpeting? Use a vacuum with a HEPA filter. The bad indoor air in general? Keep rooms ventilated with open windows and use exhaust fans in places like the kitchen, bathroom, and laundry areas. These actions are easy, inexpensive, and helpful.

Let's take a tour of some of the other egregious household pollution offenders and their sources in our everyday lives and the lives of our beloved companions. This will help you understand where you can start to pay attention and change what you clean your house with, what you spray in the garden, how you furnish and decorate your rooms, and what consumer products you bring home, including personal-care products.

Below is a partial list of common sources of household pollution, according to environmental studies and organizations like the World Health Organization (who.int), Environmental Protection Agency (epa.gov), and Environmental Working Group (ewg.org):

Aerosol sprays
Building materials (walls, floors, carpets, vinyl blinds, and furnishings)
Carbon monoxide
Cleaning materials (detergents, disinfectants, floor and furniture polish)
Dry-cleaned clothing
Heating systems or appliances
Hobby supplies (glues, adhesives, rubber cement, and permanent markers)
Insulation foam
Lawn and garden chemicals
Lead
Mold
Moth balls, moth crystals
Paint (especially those with antifungal properties)
Personal-care products
Pesticides
Plastics
Plywood, particleboard

Polyurethane, varnish

Radon

Room deodorizers, air fresheners, and scented candles

Synthetic fabrics

Tap water

Tobacco smoke

Wood preservatives

Plastic Playhouse: Life Is Stinky

Plastic is everywhere. From cars to computers, bath and pet toys to bottles, clothing to kitchen tools and storage containers, plastic is ubiquitous practically beyond measure. In the last decade we produced more plastic than during the entire twentieth century. Fully half of the plastic used in circulation is single-use, meaning it's used just once and thrown away. Most people don't realize how much the quintessential smell of plastic, especially Fido's soft, plastic chew toy, is the telltale sign of a chemical soup. The most egregious to health are the ones we've already named: BPA, PVC, phthalates, and parabens.

Although there has been a consumer-driven push to remove BPA from products, particularly those to which children are exposed (e.g., sippy cups and baby bottles), it continues to lurk, and dog toys are notoriously full of the chemical. Anything with the word "fragrance" on the label also could be problematic. According to federal law in the United States, the components of any substance labeled as "fragrance" do not need to be disclosed to the EPA, FDA, or any regulatory agency.

Interestingly, phthalates can be hidden within the "fragrance" label, as they are added to carry the fragrance and help lubricate other substances in the ingredients. Phthalates are not only found in classic plastics; they land in perfumes, hair gels, shampoos, soaps, hair sprays, body lotions, sunscreens, deodorants, nail polish, and medical devices. And they also end up in pet-care products and toys. In one of the first studies of its kind, biochemists at the New York Department of Health set out to measure exposures to twenty-one phthalate metabolites in pet cats and dogs in 2019. They recorded widespread exposure, and the levels of one of the phthalates tested were just two-fold less than what the EPA suggests is "okay" for humans.

Toxic Toys, Chews, and Beds

The following ingredients are commonly found in pet toys and chews:

- **Phthalates:** Again, this large class of chemicals often is added to pet toys made of polyvinyl chloride (PVC) to soften the vinyl and make it more flexible and gnawable. Phthalates actually smell like vinyl. Words like "methylparaben," "ethylparaben," "propylparaben," "isopropylparaben," "butylparaben," and "isobutylparaben" in the list of ingredients are clear clues, but most toys aren't labeled with their ingredients. It's not rocket science: The more dogs play and chew on vinyl or soft plastic toys, the more phthalates seep out. These toxins move freely and can be absorbed into dogs' gums and skin. The result is damage to their liver and kidneys.
- **Polyvinyl chloride (PVC):** Commonly called "vinyl," this is a relatively hard plastic but it's often filled with softeners like those phthalates. PVC also contains chlorine, so over time, as a dog chews toys made of PVC, the chlorine is released. Chlorine produces dioxins, which are dangerous, well-known pollutants. They cause cancer and immune system damage in animals. They're also associated with reproductive and developmental problems.
- **BPA:** This is the base material in polycarbonate plastics and is widely used in a variety of plastic products, including those

in your local pet store. It's also found in the lining of cans containing dog food (and in the lining of cans of human food, also). In one 2016 study from the University of Missouri, BPA was shown to upset the canine endocrine system. It also can cause disruptions to a dog's metabolism.

➤ **Lead:** We all know lead is a toxic substance, especially to the nervous and gastrointestinal systems; lead poisoning is feared by any informed individual. But people don't realize that, despite the ban on lead paint in the United States since 1978, it is *still* around. In addition to old homes that were painted decades ago, lead can enter a pet's life through imported goods like tennis balls or other toys for pets, imported ceramic food and water bowls that were glazed with lead, and lead-contaminated water.

➤ **Formaldehyde:** You probably had your first sniff of formaldehyde (and hopefully it was small) in your fifth-grade biology class. It's a long-established preservative. But it's also a known carcinogen, whether ingested, inhaled, or absorbed through the skin. It should stay locked up in those jars holding preserved specimens—not in rawhide chews, where it is prevalent.

➤ **Chromium and cadmium:** Lab tests by ConsumerAffairs put Walmart in the hot seat a few years ago, when toxicology reports revealed high levels of these chemicals in pet toys sold by the massive chain. Too much chromium damages the liver, kidneys, and nerves, and it may cause irregular heart rhythm. High cadmium levels can destroy the joints, kidneys, and lungs.

➤ **Cobalt:** Petco recalled cobalt radiation-tainted stainless steel pet bowls in 2013, spawning a new awareness of the importance of toxin-free food and water bowls and stainless steel that has been third-party checked for contaminants.

➤ **Bromine:** This flame retardant is often found in furniture foam, including the foam in dog beds. At toxic levels, bromine causes stomach upset, vomiting, constipation, loss of appetite, pancreatitis, muscle spasms, and tremors.

RECYCLE THOSE SQUEAKERS!

Dogs don't know about the toxicity of the materials that enshroud their highly sought-after squeakers. In fact, many dogs have one mission with toys: remove the squeaker as quickly as possible. If the highly rewarding act of dismembering the toy in a gleeful flurry of Poly-fil leaves you with lots of squeakers, you can recycle them into far-less-toxic DIY toys. Cover used squeakers in paper, stuff them in an old cotton sock, and knot it. Then bury the sock among crunched-up balls of crispy newspaper and watch the glee and excitement all over again (without added phthalates or PVCs!).

A Note about Flame Retardants: Chemical flame retardants are common in many products we use daily. By law, they are added to a wide variety of household items such as furniture, fabrics, electronics, appliances, mattresses, bedding, padding, cushions, couches, and carpeting. Problematically, however, flame retardants don't remain confined to the products that contain them; they migrate out of products and can contaminate house dust, which accumulates on the floor where dogs (and babies) play. As we highlighted in Part I, a 2019 Oregon State University study identified flame retardants as the likely culprit in the epidemic of hyperthyroidism among cats (the number of cats diagnosed with hyperthyroidism in 1980 was one in two hundred; today it's one in ten). Flame retardants are nearly impossible to avoid completely, but there are simple precautions you can take to minimize exposure, like adding a layer of protection in the form of an organic sheet or blanket between your dog and the sprayed surface.

A Note about Flea and Tick Treatments: These are, technically, *pesticides, right?* They prevent pests like fleas and ticks from bothering or infecting our pets. But are they toxic? If they poison pests, can't they poison pets? Warnings on many package inserts advise calling poison control if any of the product makes contact with human skin but infer that direct application to your dog's skin is entirely safe. Some of these products have come under scrutiny, alarming both veterinary associations and the EPA in recent years. A 2019 review of Bravecto and other flea and tick products containing isoxazoline, revealed *two out of three dog owners (66.6 percent) reported abnormal side effects.* Sometimes these drugs must be used, but there are safe ways to minimize the use of these potent chemicals, which help reduce chemical resistance and chemical burden in our dogs (you'll assess your dog's risks in Chapter 10). There's a strong push from environmental scientists to return to "rational use" of parasiticides in place of the overuse of broad-spectrum products that can damage animals' bodies and the environment. After all, if your dog is doused with poison to kill these bugs, anyone who plays with your dog is equally vulnerable. In addition to adopting a more conservative topical or oral chemical pesticide regimen for your dog, we'll discuss detoxification strategies in Part III.

A Note about PFAS: Per- and polyfluoroalkyl substances (PFAS) are used in a wide range of consumer products, from carpets to food wrappers and nonstick cookware. They are resistant to water, oil, and heat, and their use has expanded rapidly since their development in the mid-twentieth century. It's not surprising that these substances are ubiquitous in the environment and have been detected at high levels in dog poop. These toxins not only affect growth, learning, and behavior, but they also can interfere with the body's hormonal and immune systems and increase the risk of cancer—especially liver cancer. Our detoxification strategies outlined in Part III will help you minimize exposure to PFAS.

A Note about Air Fresheners: More than 80 percent of North Americans use some kind of air freshener—sprays, electric plug-ins, gels, and candles. But do you know what's in these products? Most people assume air fresheners are safety-tested prior to sale, but shockingly, no testing is required, and chemical companies don't need permission to sell these products to consumers for household use (fewer than 10 percent of ingredients are disclosed on the labels). Synthetic aromas are largely made up of VOCs, which float through the air and can enter your and your dog's bloodstream when the invisible particles come into contact with the skin or are inhaled. Research shows that even once-a-week use (e.g., spraying the air in a bathroom) may increase a person's odds of developing asthma and other lung diseases by as much as 71 percent.

Many of the chemicals used to formulate these fresheners—benzene, formaldehyde, styrene, and phthalates—are known carcinogens, hormone disruptors, and general irritants that can cause neurological, respiratory, and allergic responses. Most plug-ins also contain naphthalene, which causes lung cancer in animals. And studies show that **the average level of chemicals is often double in pets than people**—highlighting once again the extreme vulnerability of our cohabitating companions.

Just when you thought you could go back to old-fashioned, unscented candles, get this: The vast majority of candles are made with paraffin wax, a petroleum by-product that is created during the process of refining crude oil into gasoline. When heated, paraffin wax releases the toxins acetaldehyde, formaldehyde, toluene, benzene, and acrolein into the air, all of which increase cancer risk. Burning several paraffin wax candles at one time can exceed the EPA's standards for indoor pollution; and up to 30 percent of wicks contain heavy metals (lead), so several hours of burning foments airborne heavy metals much higher than acceptable. The number of toxic chemicals in paraffin mixtures and released through burning is dizzying (and unpronounceable): acetone, trichlorofluoromethane, carbon disulfide, 2-Butanone, trichloroethane, trichloroethene, carbon tetrachloride, tetrachloroethene, chlorobenzene, ethylbenzene, styrene, xylene, phenol, cresol, cyclopentene. We won't bother to define these substances worthy of chemistry credits.

Solution: Don't buy any product that lists "fragrance" on the label or "made with" as a marketing tool. Replace paraffin candles with unscented candles made with 100 percent beeswax, soy, or vegetable wax. Check a new candle for lead wicks by rubbing the wick on a piece of paper; if it leaves a gray pencil mark, the wick contains a lead core. You can water diffuse dog-friendly pure essential oils from reputable companies in one room of your home, always leaving an escape route to an area where your pets can retreat that has no added natural aromas whatsoever. Simmer orange peels and cinnamon sticks on your stove. And (all together now): Open the windows!

Air quality can be impacted by more than VOCs from scenting products and off-gassing. Forest fires, city pollution, smog, secondhand smoke, and mycotoxins from water-damaged homes can all impact your animal's respiratory and systemic health as much as your own. Identifying and removing sources of poor air quality is important; it's as simple as getting an indoor air purifier if you live in a city, where air quality may be poor, or performing mycotoxin testing if your home has had water damage.

How Bad Is Your Water?

Unfortunately, this question cannot be answered easily; the look and taste of your water are inadequate indicators of its quality. You may splurge on pristine water for yourself via a water company or filtration system in your kitchen or refrigerator, but what is Fido drinking? Numerous toxins may be present in your tap water. Thankfully, your community's Annual Quality Report addresses the safety of your public water supply. To find this report, check the NRDC (Natural Resource Defense Council) report "What's on Tap?" at www .nrdc.org, or ask your water utility (the company that gives you a monthly bill) for a copy of the their annual results. Found in the report will be a list of contaminants, the potential source(s), and how much of the contaminants were found in the water.

We all heard about the horrific water crisis in Flint, Michigan, that started in 2014 and that poisoned the water supply with lead. But often the presence of contaminated water is far less obvious. In 2020, researchers at the University of Illinois at Urbana-Champaign published a paper highlighting the problem of "anthropogenic contaminants" in water; these are the pollutants that end up in our water supply from our own human behavior—runoff from farming and livestock activity, disinfecting techniques, and therapeutic drugs released into sewage. The paper was particularly critical of the ease with which EDCs (endocrine-disrupting chemicals or "xenoestrogens," environmental chemicals that act like hormones in the body) infiltrate the water supply, harming both humans and nonhumans (ahem, our dogs). Microplastics, heavy metals, and chemical contaminants in drinking water should be removed before anyone in your family, including your animals, drinks tap water.

We highly recommend using filtration systems in your home. Some filters—like pitcher filters—require you to manually fill them, while others attach to your plumbing. These include under-the-sink filters and faucet attachments. The function of a particular filter could be to create clearer, better-tasting water or to remove toxins. A majority of filters do both jobs. Depending on the design and filter media in the unit, filters can reduce many types of contaminants, including chlorine, chlorination by-products, lead, viruses, bacteria, and parasites.

Shoes with Strings Attached

As we'll recommend in Part III, one of the easiest, cheapest ways to free your house of indoor toxins is to take your shoes off before you enter it. This can limit exposure to harmful substances like lawn chemicals, carcinogens found in asphalt and petroleum byproducts, fecal matter on sidewalks, and any sort of bacteria, virus, allergen, or toxin. It doesn't take much imagination to think about what you trudge through out in the real world and then bring into your home on the bottoms of your shoes. Even a fancy pair of Manolos, Tom Fords, or Nike Airs will carry invisible toxins into the room with you. In fact, your shoes may be even more toxic than your toilet! So when you catch your dog lapping up toilet water, think about what else Fido is licking up from the floor when you drop scraps of food while cooking dinner.

Can chemicals in the environment cause weight gain?

In 2006, Dr. Bruce Blumberg of the University of California, Irvine, created the term "obseogens" to classify the chemicals that make us fat. Blumberg found that mice he was using in one unrelated chemical study were gaining weight, so he began researching the connection between fat and chemicals. That made him wonder whether there is an alternative explanation for our persistent inability to lose weight. Research confirms his suspicions. Since then, numerous studies in both humans and animals have identified a strong link between exposure to certain environmental chemicals and greater body mass index (BMI).

Obesogens interfere with the formation and balance of fat metabolism, or how your body develops and stores fat. They do this by turning stem cells into fat cells. Additionally, obesogens can alter how your body responds to the calories and nutrition you take in. And many of them can do a number on our hormone system. What's perhaps most destructive about obesogens is that they may be passed down.

That's right: The effects of obesogen exposure can be *heritable*, largely through epigenetic forces. The consequences of obesogens may be felt by any and all of your descendants. The science of obesogens is complex, but suffice it to say they are pervasive in our everyday lives, and many of the same chemicals we've covered qualify as obesogens (e.g., chemical pesticides, BPA, PFOAs, phthalates, PCBs, PBDEs, parabens, and air pollutants).

Noise, Light, and Electrostatic Pollution

A smoggy skyline is an obvious sign of air pollution. But we neglect to consider other forms of pollution that stealthily inhabit our lives: excess noise and light. These are the necessary evils of modernity. They reflect our accomplishments as a civilization, but they come with liabilities—namely, disruptions to our natural rhythms that prefer to follow the twenty-four-hour solar day. Put simply, too much noise and light, particularly at times when the body needs darkness and silence, is damaging to health. Light pollution is an old problem, from records dating back to the late 1800s that describe migrating birds flying into lighthouses. But light pollution has intensified in the past century, and the level of noise to which we are exposed daily now far exceeds that of previous generations. Noise pollution needn't be supremely loud to be debilitating. The drone of televisions (and other screens) and the whir of typical urban life (sirens, lawn mowers and blowers, garbage disposals, thunder, aircraft) disrupt our body's natural rhythms.

Noise pollution has come under scrutiny lately in scientific circles. Recent studies show that people who live near airports are at increased risk for cardiovascular disease—separate from any heightened cardiovascular risks associated with the air pollution. One study published in the *BMJ* found that people who lived in the noisiest areas (i.e., near an airport) had an elevated risk for stroke, coronary heart disease, and cardiovascular disease, even after adjusting for confounding factors such as ethnicity, social deprivation, smoking, road traffic noise exposure, and air pollution. In addition, the body's biological response to noise was dose-dependent; the risk was greatest in the 2 percent of the population who experienced the highest levels of noise.

Sound is a form of electromagnetic radiation whose pitch or frequency is measured in hertz (Hz). A Hz is defined as one complete wave cycle per second. The difference between human and dog frequencies is 20 to 20,000 Hz for humans and 40 to 45,000 Hz for dogs, respectively. Additionally, cats can hear the highest frequencies of the three up to 64,000 Hz. Therefore, cats and dogs can hear from farther away than humans can. Furthermore, the intensity or loudness of a sound is measured by decibels (dB). After 100 dB, hearing damage is automatic.

While you might think the health impairment from exposure to constant or high levels of noise results from disrupted sleep brought on by the noise, it turns out the connection is much more direct. Chronic noise leads to continual stress on the body, which in turn causes higher blood pressure and elevated heart rate, endocrine disruption through the stress hormone cortisol, and heightened overall inflammation. Whether these studies translate to canine health concerns is currently under investigation, but we suspect these heightened risks are borne by our furry friends, too, especially since dogs are wired to assess their environment through auditory cues unmuddled by the unnatural and all-consuming acoustics of thumping subwoofers, surround-sound news broadcasts, and constant radio chitchat. We know, for example, that dogs can often develop a sensitivity to loudness, different pitches, or sudden noises, and these sensitivities can manifest behavioral issues like abnormal fear and anxiety. The relationship between noise sensitivity and fear has long been documented in dogs, regardless of breed, though variations do exist based on breed, age, and sex (older female dogs are at highest risk for heightened sensitivity).

In one 2018 study, a group of animal behavioral scientists from the UK and Brazil revealed a link between noise and underlying physical pain. The researchers suggest that pain, which could be undiagnosed, is exacerbated when a noise makes dogs tense up, putting extra stress on their muscles or joints (which are already inflamed, causing further pain). That pain is then associated with a loud or startling noise, leading to a sensitivity to noise and avoidance of situations where they previously had a bad experience, such as a hostile encounter at a local park or a loud room in the house. This study also means that sensitivity to noise could very well be a cry for pain relief.

The outer-ear anatomy, otherwise known as the pinna, on dogs, cats, and horses is what enables them to be more sensitive to sound than humans. Many species, including laboratory animals, can experience hearing loss or noise-induced stress. Even weeks after the sound environment has returned to normal levels, loud sounds or chronic noise disruption may cause you or your pet to experience elevated blood pressure. One study showed that dogs can also experience negative side effects from noise, such as increased heart rate and salivary cortisol levels. If ambient level of 85 dBs persist on a consistent basis, canines can experience anxiety. To measure hearing loss in pets, a test called the "brain auditory evoked response" (BAER) was conducted on dogs in a kennel where background noise consistently reached 100 dB. Within six months, all fourteen dogs had some degree of hearing loss. We can only imagine what that noise exposure did to those dogs' baseline levels of fear and anxiety.

Noise phobias in dogs can have a genetic, hormonal, and early socialization component. It can take noise-sensitive dogs four times longer to calm down after stressful situations, probably chucking impressive amounts of stress hormones the entire time. Animal research also demonstrates behavioral changes when exposed to extremely low levels of electromagnetic fields (EMFs). We recommend creating a noise-muffling, EMF-free zone that's also free from "junk lighting." We'll show you how to do this in Part III. (Hint: Turn off all continual sources of ongoing noise and sources of EMFs every night, including the TV, computers, and your router.)

On the light front, studies of shift workers have revealed how exposure to light at the wrong time of day can do a number on the body. People who work night shifts may think they can "train" their bodies to stay awake at night and sleep during the day, but the research tells a different story. Shift work has been linked to obesity, heart attack, several types of cancer (breast, prostate), a higher rate of early death, and even lower brainpower. And the connection has everything to do with the relationship between light and our circadian rhythms. Dr. Satchidananda Panda, whom you met earlier, has worked extensively on the circadian clock of humans and animals, especially in relation to genes, the microbiome, sleeping and eating patterns, risk for weight gain, and the immune system.

One of his most important discoveries demonstrated how light sensors in the eyes work to keep the rest of the body on schedule. The suprachiasmatic nuclei in the hypothalamus, that special part of the brain tied to emotions and stress, is the seat of the biological clock in all mammals. It receives input directly from the retina in the eyes and serves to "reset" the circadian clock. This is why exposure to early-morning light helps reset the circadian clock and why getting out into the morning sun can help to recalibrate your clock.

Dr. Panda believes pets stuck in homes all day with closed drapes develop depression because their brain can't create and secrete the appropriate neurochemicals for healthy synapses. His research shows that animals' physiology is regulated in part by the light signals that directly enter their eyes. These light signals trigger a series of chemical signals inside the brain and, in turn, the body. When dogs are outside in the morning, the light signals the brain to release melanopsin—a light-sensitive protein—and to wake up. Later in the day, when dogs go outside at dusk, the light signals the brain to release melatonin—the "sleep" hormone—and to prepare for sleep. According to Panda: "The eyes have specialized cells— 'blue light-sensing melanopsin neurons'—that connect to parts of the brain involved in depression, feeling happy, and melatonin production. Experiments show when animals don't have these blue light sensors activating during the day, they feel depressed."

Similarly, dogs who are overexposed to bright, synthetic lighting also suffer. According to Dr. Panda: "Even healthy lab mice put under constant light for three to four days will become medically sick. If you look at their blood, cortisol, inflammation levels, hormones— everything is out of whack." Dr. Panda further notes that these animals become more glucose intolerant, and the early signs of diabetes show up very quickly. So this isn't just about mood and behavior; it's also about managing the metabolism and immune function.

Panda believes it's our responsibility, as caretakers, to enable our dogs to regulate their own clocks by having direct access to the outdoors at least twice a day. We recommend combining these important, light-sensing circadian excursions with "sniffaris," Dr. Alexandra Horowitz's recommendation to allow dogs to sniff to their hearts' content at least once a day. Our recommendation is to **host a circadian-setting sniffari for your dog for a few minutes in the morning and again in the evening before you go to bed**. These walks aren't for cardio, they're for brain health, providing circadian regulation, neurochemical regulation, and olfactory stimulation that enhances cognitive well-being.

Every cell of the mammalian body expresses a circadian clock, driving 5 to 20 percent of genes to be expressed with a twenty-four-hour, night/day rhythm that hinges—physically—on sleep habits. This rhythm consequently drives daily timing of many aspects of behavior and physiology. Examples of daily rhythms in physiology include processes such as blood-sugar balance, hormone release, and immune response. Behavioral rhythms include sleep-wake patterns as well as timing of eating, pooping, and exercise. The daily timing of these behaviors and physiology has evolved to be able to anticipate and prepare for changes in the environment such as food availability and the light-dark cycle. Disrupting these biological rhythms increases the risk of developing health issues, including diabetes, obesity, and cancer.

Dogs and humans possess different circadian rhythms and sleep patterns, but the rules remain the same: We each need to get sufficient nighttime sleep and to follow certain patterns that keep our rhythms in check, which in turn impact everything else about us—from the flow of hormones to metabolism to immune function. Dysfunctional relationships with light exposure and poor sleep habits can have enormous biological consequences we're just beginning to understand in science and medicine.

A muddy dog—one that loves to roll in the dirt and run through open green pastures touched only by nature—is not the same as a dirty dog—one laden with the burdening residues of modern life. Now let's get to the How-Tos. With an understanding of the science, making sense of the solutions is easy.

LONGEVITY JUNKIE TAKEAWAYS

- ➤ We live in a sea of constant exposures, and our pets often bear more of the "body burden" brunt because they are closer to the ground and cannot take many of the same kinds of precautions we can to reduce or mitigate those exposures.

- ➤ Studies reveal vast amounts of chemical residues in the urine of both cats and dogs exceed healthy thresholds. Reduce chemical exposure by "greening up" your home, inside and out.

➤ These chemicals are found in obvious places, like cleaning agents and herbicides in lawn care. But they're also found in unlikely places, such as scented candles and air fresheners, off-gassing furniture (including the pet bed!), the water supply, flea and tick treatments, and all manner of plastics, including those in popular pet toys.

➤ Dogs are especially sensitive to excess noise, light at the wrong time of day, and electrostatic pollution. Overexposure to artificial light adversely impacts a dog's metabolism, immune function, mood, and behavior. Open up your shades in the morning, and turn off the lights, computer, router, and TV at night.

➤ Twice-daily sniffaris—one in the early morning and another in the evening—are great ways to reset your dog's biological, circadian clock.

➤ To sum up Parts I and II, and in preparation for Part III, it helps to think about the 5 R's every day:

1. **Reduce** processed food.
2. **Revise** meal timing and frequency.
3. **Ramp** up physical exercise.
4. **Refill** deficits and deficiencies with supplements.
5. **Rethink** environmental impacts (stress, exposure to toxins).

Let's get to it!

PART III

Pooch Parenting to Build a Forever Dog

7

Dietary Habits for a Long and Healthy Life

Building a Better Bowl

Food and medicine come from the same source.

—Chinese proverb

Christmas Eve 2020 yielded a wonderful and unexpected blessing for the Becker household: Homer. We got word a twelve-year-old Glen of Imaal Terrier was homeless after his dad passed away at a local nursing home. Homer wasn't homeless very long. Although Homer said no to all the raw veggie bites I offered him on Christmas Eve, he said yes to a bite of steamed carrot and a sliver of apple on Christmas Day. He's been trying new foods every day since then. He's now eagerly accepting the same foods he declined a few months ago, expanding his palate, microbiome, and nutrient intake. We slowly weaned him off his ultra-processed dog food for "sensitive stomachs," and now he enjoys an array of real-food meals that have positively impacted his health: his dull, dry coat is being replaced with a lustrous shiny coat, he's shedding much less, his gas has resolved, he's lost his pudgy midsection, his breath is better, and he's moving better (he's less stiff and has better endurance).

These poignant stories of restoration and rejuvenation are

commonplace among Longevity Junkies, and for one reason: Fresh food changes dogs in *amazing* ways. Twenty-five years later, I still get giddy witnessing these transformative metamorphoses. It's soul satisfying to snuggle with Homer and know that his precious life is better today—and almost certainly extended—because of the simple but incredible power of *real food*. Very simply, it's the best gift we can give our dogs.

Which of the following two journeys would you like to take with your dog?

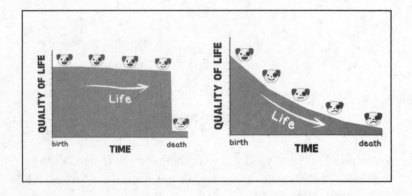

In all likelihood, you chose the option on the left, which means you're happy throughout your life and then one day you go *poof!* in the night during sound sleep—like the clock ran out but ticked long and hard until it stopped. Sure, you have some ups and downs and the usual disappointments and triumphs of life, but there's no life-sucking, soul-crushing physical degeneration over time, no loss of cognition or mobility, and no erosion to quality of life despite the passage of years. You live robustly until the last few breaths. Wouldn't that be grand? And wouldn't you want that for your dog? It's why we wrote this book.

Welcome to the actionable part. You may be thinking: *Now tell me what to do!* Go ahead and pat yourself on the back first. You've gained a tremendous amount of science and knowledge by now, and this next section is putting all of that information to work. It's not important that you remember all the details, but rather that you have the foundational understanding of why and how making some changes will improve the overall health (and happiness!) of your dog. As important, we will guide you in making changes that work for your specific circumstances, time, budget, and inclination. You've probably learned more about the habits of a highly effective mammalian body than most of today's doctors and vets. If you haven't already begun to change a few things in your life based on what you've read, now is your chance.

When you put these suggestions into practice, you'll increase the likelihood that you and your beloved dog will bask in a high quality of life for as long as possible. We're serving up a big portion of How-Tos with a side of scientific reminders to explain why. And that's important because it's challenging to meaningfully anchor behavior change in lifestyle habits without knowing *why*. When you understand the *why*, following the *how* becomes straightforward, enjoyable, and rewarding. **The goal is aspirational, indeed, but achievable: to parent a Forever Dog that lives as vibrantly and happily as possible to the finish line**. And in doing so, you as a devoted pet parent reap many health rewards as well, from purely physical improvements measurable by a doctor to those intangible but invaluable benefits like enhanced confidence and self-esteem, and feeling more youthful and in control of your life and future. In short, you will be healthier, happier, and more productive. And your success will breed more success; when you start to see and enjoy their benefit from even simple changes, you may well decide to up your game. We know you can do this for yourself and your pet. The payoffs are huge for everyone. Most important, you don't have to implement all of our suggestions at once or even ever. Start with what's easiest and makes the most sense for you.

Our social media conversations about superfoods and supplements are among the most popular. In a post about Scottish Terriers, for example, we shared the data indicating that consuming any type of vegetable at least three times a week was associated with a 70 percent reduced risk for developing transitional cell carcinoma, otherwise known as TCC, a type of cancer most commonly found in the urinary bladder and urethra of older small dog breeds (such as Scottish Terriers). Consuming yellow-orange vegetables and green leafy vegetables at least three times as week is associated with an approximately 70 percent and 90 percent reduction in risk of developing TCC, respectively. Pet mommas and papas crave this information, and we feel honored to be in a position to share life-changing nuggets of science to our wellness community.

Most important, adding these simple and easy Longevity Foods has a profound impact on your dog's health. The best part of "bowl additions" means you add these foods to whatever you're currently feeding; you don't have to completely overhaul your dog's entire life right now to achieve notable improvements in well-being. **Small amounts of Longevity Foods hyper-fortify your dog's current meals with powerful antiaging nutrients and cofactors**. Every small change is a step in the direction of optimal health and longevity.

This book may be your first introduction to the idea of bringing your dog into your own personal health and wellness journey. Many health buffs we know feel ashamed or embarrassed that they never thought about the amount of "fast food" they were serving their dogs until they caught one of our Facebook Lives. Then they felt frantic, trying to incorporate everything they felt they should be doing, and ended up with sheer anxiety about what they might be forgetting.

Many 2.0 pet parents (and human scientists and researchers) who are familiar with the longevity principles and science outlined in Parts I and II are now searching for direction on how to institute a comprehensive canine lifestyle overhaul. Our goal is to provide every dog owner with a full toolbox of options, so you can make lifestyle changes à la carte—at a pace that makes sense for you (and your dog). We do this through our simplified plan: the Forever DOGS Formula:

- ➤ **D**iet and nutrition
- ➤ **O**ptimal movement
- ➤ **G**enetic predispositions
- ➤ **S**tress and environment

We'll present lots of lists and ideas (e.g., Longevity Foods you can share with your dog, foods you should never feed, safe herbal remedies that act as antianxiety pills, navigating the supplement aisle), and help you tailor the plan to your own life and dog. Think of it as a customizable lifestyle plan for your pooch! We'll start with what we feel is the most important place to begin, no matter what your health goals are for your pet: diet and nutrition from real food.

Diet and Nutrition

We've made it clear by now that **all good changes to extend life start with food**. Transformations in health that ultimately improve one's experience on this planet and the timing of one's demise hinge upon diet and nutrition, whether you're a dog or a human. Through good nutrition, we sustain ideal weight; nourish the microbiome; support metabolism, detoxification, and overall biology. These factor into everything health-wise; and increase our motivation to do more to maintain optimal health, such as get a good night's sleep, exercise and physical fitness, stress and anxiety management, and enduring the inevitable challenges that come with life, including unwanted exposures to things that can have adverse effects.

The best diet for your dog will depend on a few variables, such as her age and health status and any preexisting conditions. We are huge proponents of more fresh foods for pooches. What does that mean?

Incrementally Improving the Nutrition in Your Dog's Bowl

People are limited by time and money when it comes to feeding their furry friends. You want the best for your dog, but realities must also be considered. Research shows 87 percent of people already are adding other foods into their dog's bowl. We love this, but we think we can capitalize on these good intentions by ensuring that key Longevity Foods also make it into their bowls.

In the next chapter, we'll provide you with criteria for evaluating dog food, both the food you're currently feeding as well as any fresher brands you are considering buying. If you find yourself wanting to upgrade to fresher brands and food choices, it doesn't have to be an all-or-nothing approach. There are endless ways to spruce up your dog's bowl, but to make things easy, we've broken down improving your dog's nutrition into two steps: (1) introducing Longevity Foods, and (2) evaluating and altering your dog's daily diet, if needed, after you've completed the Pet Food Homework we provide (don't worry, it's not algebra).

Introducing Longevity Foods

The starting point we recommend for everyone, regardless of what you're feeding your dog now or will be feeding your dog in the future, is to strive to make 10 percent of your dog's calories come from Longevity Foods. We didn't arbitrarily choose 10 percent; we chose 10 percent because no veterinarian or veterinary nutritionist will object to this incremental change. Vets have **the 10 percent rule: 10 percent of your dog's calories can come from nutritionally incomplete foods and not "rock the balance."** Many people burn through their 10 percent (add-ins or extra) freebie calories in wasteful (sometimes shameful) ways, feeding poor-quality starchy, carby, truly junk-food treats. We're hoping to persuade you, right now, to maximize the health benefits of your dog's daily

10 percent caloric allotment by swapping those junky, poor-quality, biologically stressful treats (that provide no nutritional value and adversely impact your dog's health) with Longevity Foods.

We're not saying true treats don't matter, but we do want you to rethink the types of treats you provide, along with when and why. We'll cover the timing of treats a little later, when we discuss your dog's circadian rhythm, but right now we want you to think about reframing the concept of treats. For starters, we want you to stop thinking of treats as *snacks*; rather, view treats as foods that *medically treat* the body, that *nourish* your dog's cells. Imagine providing the body with such stellar Longevity Foods that these new, healthy treats actually cultivate healthy organ function, the microbiome, the brain, and the epigenome. This happens by substituting industrially produced, ultra-processed snacks for Longevity Foods. It's similar to what we humans do when we make our own snacks—we stop munching on candy in the afternoon and turn instead to a handful of nuts or veggie sticks with homemade guacamole. Small shifts can have big effects.

We get it, your dog may look at you like you're taking this new health kick a little too far. You may be dreading the soul-piercing "Where's my Pup-Peroni?!" stare, but it's worth noting that 10 percent of your dog's daily calories used "incorrectly" can actually undo some of your longevity goals. We've discovered many health-conscious humans regard treats as being a nonfactor in their dog's health, but just as we can undo the healthiest of meals by finishing with a slice of chocolate cake, even the best of health regimens are compromised by crappy, carby pet treats. *We want to encourage you to use part or all of your dog's free 10 percent of calories a day for good: to fortify and nourish your dog on a cellular level.*

Thankfully, most of the Longevity Foods (especially the fresh veggies and fruits) are so low in calories, they really don't even count, but even small bites are big scores for health. Longevity Foods are nutrient powerhouses, so you don't have to feed huge amounts to realize huge health benefits. Best of all, they can be fed as treats or added directly to your dog's bowl, so we also refer to them as *toppers* (they "top off" whatever is in the bowl). If you don't routinely give your dog treats, then you'll be using Longevity Foods as Core Longevity Toppers (CLTs), mixing them right into whatever food you're currently feeding your dog.

Some Longevity Foods are best used as toppers versus healthy treats because they're too hard to use as training treats (like sprouts—they're stringy and can be messy if you put them in a treat bag). We've made lists of the Longevity Foods that are easiest to feed as treats (see page 246). You can cut all of these suggestions into pea-size morsels and dole them out as lures/rewards throughout the day. Yes, you heard us right: Whether you have a mini Aussie or a Cane Corso, we recommend pea-size "treats" (Longevity Foods–turned–training rewards); bigger dogs simply get more of them. If you have a finicky dog used to eating junk food for treats, you'll have a hard time convincing her to eat one quarter of a brussels sprout right off the bat, so start with gently braised or poached micro-morsels of organ meats (again, the literal size of a pea; see the list on page 233 of the wide variety of nutrient-dense organ meats dogs *love*). Most dogs won't turn down cooked liver or chicken hearts. Next time you cook organ meat treats, throw in some carrots, as a lot of *super-finicky* dogs love chicken-flavored carrots. Over time, cook the veggies less and less, until she acclimates to eating raw carrots.

If you aren't home during the day, you aren't there to feel "treat pressure," so you can add Longevity Foods right on top of your dog's current food as a topper. (If you have a finicky dog, mix the toppers directly in with the food, hiding the new "health foods" within her usual diet.) The great thing about the 10 percent rule is these add-ins don't have to be nutritionally balanced—they're "extras" that work longevity magic. If your dog turns her nose up at what you've chosen to offer that day, don't despair: Use a smaller amount of a blander Longevity Food at the next meal, and chop the goodies up *really fine*; adding Longevity Junkie Homemade Bone Broth (see page 242) also helps persuade dogs to try new tastes. Sometimes it takes months for older taste buds to wake up, but keep at it. And keep notes in your Forever Dog Life Log, a journal we recommend to track your dog's experiences, likes/dislikes, and health issues. It can be an old-fashioned notebook or a digital file you keep on a computer. It also helps to write down changes in your dog's daily life, such as when a heartworm pill was given, what day diarrhea started, when he started a new food or supplement, and so on.

Longevity Foods pack powerful punches in reducing oxidative stress and positively influencing the epigenome. This, in turn, ultimately influences how your dog's underlying DNA behaves. CLTs (bowl add-ins) tip the scale on a daily basis, providing your dog's body with a steady stream of free radical–quenching antioxidants, age-extending polyphenols, beneficial phytochemicals, and key co-factors that are ready to be passed up the food chain and whisper healthful encouragement to your dog's epigenome.

Paws for Thought: If your dog is pudgy and needs to lose a little weight, you can *replace* 10 percent of his food with Core Longevity Toppers (take away 10 percent of her dog food calories and replace them with Longevity Foods). If your dog is lean, you can *add* them on top of her typical food portions.

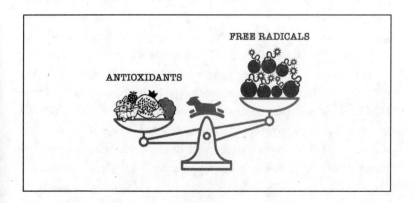

10 Percent Core Longevity Toppers (CLTs): Longevity Foods are added in the form of CLTs to whatever brand of food you're currently feeding your dog (or will be feeding your dog in the future). As we've explained, veterinarians worldwide agree that 10 percent of your dog's daily calories can come from foods other than his nutritionally complete and balanced "dog food" (i.e., through treats or other foods). We have reworked the widely accepted "10 percent treat rule" to your dog's health advantage by replacing ultra-processed treats with CLTs. We call this the 10 Percent Core Topper Rule: Replace current ultra-processed treats that don't offer any real health benefits with supercharged Longevity Foods.

There are endless ways to customize meal plans and bowls. Don't get too hung up on exact food percentages or having to make any decisions right now. There is no one-size-fits-all approach to feeding your dog the right kind of nutrition, and you can alter your food brands, percentages, and amount according to what's best for you and your pet. At this point we merely want you to be thinking about what a Forever Dog Meal Plan would look like for your dog's bowl. This means you aren't switching what your dog is eating for dinner tonight. Instead, you'll begin diversifying the bowl with Longevity Foods. In the next chapter, you'll learn how to evaluate your dog's base diet and, if needed, improve the quality, biological appropriateness, freshness, and nutrient value by choosing different categories or brands of food. Right now, we want you to think about your nutrition goals for your pup, after all you've learned in the first two sections. We hope you feel reassured by the science and understand deep down that fresh foods are best for your pet.

This can be very reassuring, especially when people say, "You take better care of your dog than most people do of themselves." It's true that creating exceptional health is hard work; there isn't a pill we can give our dogs to create vibrant bodies and double their life span. Every decision we make for our dogs has a health outcome, good or bad, and 2.0 dog owners recognize they have a finite window to intentionally create well-being. We also recognize the very definition of health is different for everyone; we want to meet you where you are and offer practical tips to improve your dog's well-being.

Core Longevity Toppers (CLTs): Superfoods You Can Share with Your Dog on a Daily Basis

There's a long list of foods with impressive longevity benefits that can be mixed in with your dog's current food or used as treats, supercharging her nutritional status in all sorts of health-promoting ways.

Fresh vegetables and low-sugar fruits are extremely important for dogs, even though they should compose only a small percentage of their diets. In the wild, wolves and coyotes consume grasses, berries, and wild fruits and vegetables as sources of crucial nutrients, which provide not only roughage (fiber) but also a variety of nutritive substances not found in meats, bones, and organs. Research shows dog diets without adequate vegetable matter create less healthy microbiomes. Among the most important compounds plants provide are the polyphenols, flavonoids, and other phytonutrients. In multiple studies, adding polyphenols to the diet has been shown to significantly reduce markers of oxidative stress. They are found in an abundance of dietary sources.

We humans get a good dose of polyphenols from coffee and wine, but we obviously don't recommend sharing your morning joe and dinner vino with Fido. For us, coffee and wine in moderation are common micro-sources of these antiaging molecules (for many, coffee is the only source of dietary antioxidants in a day). But the food sources, in the following chart, are all dog-friendly foods you can sprinkle on her food or share with her as you're enjoying them.

Although the volume of roughage (veggies) in biologically appropriate diets is relatively small for dogs, their inclusion plays a

Types of Polyphenols

Classification		Representative Members	Food Sources
flavonoids	anthocyanins	delphinidin, pelargonidin, cyanidin, malvidin	berries, cherries, plums, pomegranates
	flavanols	epicatechin, epigallocatechin, EGCG, procyanidins	apples, pears, tea
	flavanones	hesperidin, naringenin	citrus fruits
	flavones	apigenin, chrysin, luteolin,	parsley, celery, orange, tea, honey, spices
	flavonols	quercetin, kaempferol, myricetin, isorhamnetin, galangin	berries, apples, broccoli, beans, tea
phenolic acids	hydroxybenoic acid	ellagic acid, gallic acid	pomegranate, berries, walnuts, green tea
lignans		sesamin, secoisolariciresinol diglucoside	flaxseeds, sesame
stilbenes		resveratrol, pterostilbene, piceatannol	berries

critical role in repairing and maintaining digestive and microbiome health. Veggies provide prebiotic fibers for short-chain fatty acid production in the colon. They also provide the soluble and insoluble fiber necessary to maintain healthy elimination and immune-boosting, antioxidant-promoting phytonutrients.

The following list highlights just a few examples of the veggies and fruits you may have in your refrigerator right now that can add valuable nutrition to your pet's diet when fed as CLTs to their foundational meal or as training treats throughout the day. **CLTs are morsels of fresh food you can feed raw or gently cooked (steamed is a smart choice for both of you, if you're cooking).** Feel free to recycle a few of the cooked veggies you made for your dinner last night into your dog's bowl this morning (just make sure there are no sauces that could cause GI upset). You can finely chop any of these dog-friendly human foods and mix them into your dog's meals, or use slightly larger morsels as training treats; either way, your dog will be consuming more living, whole fresh foods. Take a good look at the pieces and parts of the veggies you're throwing out: The tops and bottoms of the carrots, celery, green beans, and other dog-safe veggies can be chopped up and added to your dog's bowl. All fresh foods you offer your dog (cooked or raw) should be chopped up into tiny, bite-size morsels. Offer one piece at a time, and note in your dog's Life Log which foods are an instant hit and which are a "try again."

As we go through this list, we'll let you know what volume of goodies we routinely use for thirty-pound Shubie, Rodney's nine-year-old Norwegian Lundehund mix. We encourage you to try tiny bites of lots of real foods, rather than burn up your 10 percent calorie freebies on one large fresh food item. Remember these are superfoods, so they don't need to be fed in large quantities to achieve big results. In most cases, it would be hard to overdose your dog on celery (unless you have a Lab or a Golden Retriever without an "off" switch). Most of these ultra-low-calorie goodies can be fed without having to calculate calories, and the exceptions are noted. Our goal is to provide incredible fresh food diversity to build the microbiome and bolster intracellular nutrients, antioxidants, and polyphenols. Try to buy organic or spray-free whenever possible.

Longevity Veggies

Some apiaceous vegetables (e.g., carrots, cilantro, parsnips, fennel, celery, parsley): These gems contain polyacetylenes, an unusual class of organic compounds that has antibacterial, antifungal, and antimycobacterial benefits. They play a key role in detoxifying several cancer-causing substances, specifically mycotoxins (including aflatoxin B1). Mycotoxin contamination in feed-grade pet food poses serious health risks, and eliminating mycotoxins once your dog eats them can be difficult. Serving up these veggies is a great way to enhance the metabolism of these toxic compounds. Raw or cooked, organic carrots and parsnip slices make great training treats, while cilantro, parsley, and fennel can be minced and mixed in with your dog's food. Research demonstrates cilantro has a synergistic effect with chlorella for heavy metal detoxification (also a problem in the commercial pet food industry), naturally binding an average of *87 percent of lead, 91 percent of mercury, and 74 percent of aluminum* within 45 days!

Brussels sprouts: Cancer research has found cruciferous vegetables, including brussels sprouts, to have positive influences on bladder cancer, breast cancer, colorectal cancer, gastric cancer, lung cancer, pancreatic cancer, prostate cancer, and renal cell carcinoma, in part due to a bioactive compound called "indole-3-carbinol." In addition to providing gut-building fiber, brussels sprouts also contain flavonoids, lignans, and chlorophyll and are a good source of vitamin K, vitamin C, and folate as well as selenium. Most dogs prefer these to be gently cooked/steamed.

Cucumbers: Mostly water and calorie-free, these crunchy snacks are great for keeping your pup hydrated and providing vitamins C and K. They also contain an antioxidant called "cucurbitacin," which has been shown to inhibit the activity of cyclooxygenase-2 (COX-2), a well-studied pro-inflammatory enzyme, and to induce apoptosis in laboratory studies. They also contain pectin, a naturally occurring soluble fiber that benefits the microbiome.

Spinach: This green leafy vegetable has anti-inflammatory properties and supports heart health (thank you, vitamin K). The phytochemicals in spinach reduce hunger for simple sugars and fats. Spinach is the richest vegetable source of lutein and zeaxanthin, which prevents age-related eye degeneration in animal models, and it also contains alpha-lipoic acid, an important antiaging antioxidant, and folate, an essential B vitamin that helps with the production of DNA. Without folate, new and healthy DNA cannot be created. Cell biologist and longevity researcher Rhonda Patrick, PhD, maintains that "a deficiency in folate is equivalent to standing under ionizing radiation due to the DNA damage it causes." Recent research has also revealed that folate has a protective effect on telomeres, which are structures at the end of chromosomes that can be damaged by ultra-processed foods and other stressors.

As we mentioned before, telomeres become shorter over time. This decrease is linked to a shorter life span and higher risk for disease. Being highly heat-sensitive, folate is one of the first nutrients to be inactivated in processed pet food. This veggie is naturally high in oxalates, which can be a problem for some dogs genetically prone to oxalate bladder stones. We're able to hide a tablespoon of minced spinach in Shubie's bowl a couple of times a week. Being a food connoisseur, she prefers steamed spinach, still slightly warm, with a dash of paprika and a spritz of lemon (aka Rodney's leftovers).

Broccoli sprouts: Dr. Patrick notes that broccoli sprouts are *"very* powerful for antiaging." This is a good thing because, in the modern age, we face an onslaught of toxins, from benzenes (which we inhale from exhaust) to pesticides (which we ingest when we eat nonorganic food). These toxins damage our mitochondria and cause inflammation throughout the body, which speeds up the aging process. One of the body's stress-response pathways (Nrf2, or nuclear factor erythroid 2-related factor 2) controls over two hundred genes responsible for anti-inflammatory and antioxidant processes. Activation of this pathway staves off inflammation, jump starts detoxification, and allows antioxidants to do their job.

So what about broccoli sprouts? Cruciferous vegetables, including broccoli, broccoli sprouts, and brussels sprouts, contain sulforaphanes, a compound with an unrivaled ability to trigger the Nrf2 pathway. Sulforaphanes have been shown to slow the rate of cancer and cardiovascular biomarkers, decrease inflammation markers, and remove toxins (such as AGEs, heavy metals, and mycotoxins) from the body in both animals and humans. **Sprouts are the best way to remove AGEs from your dog's body!** This is an inexpensive and powerful way to flush out the toxic by-products in the ultra-processed foods our dogs consume. The sprouts are biologically superior to the "adult" vegetable because they contain fifty to one hundred times more sulforaphanes than mature broccoli and other cruciferous vegetables. If they are hard to find at your local grocery store, you can easily sprout them yourself. Hide a pinch of these amazing little gems in your dog's meals for every ten pounds of body weight.

Mushrooms: In addition to being a natural source of gut-nourishing prebiotic fibers, mushrooms contain a variety of longevity-promoting substances including polyphenols, glutathi-

How to Grow
Sprouts for You & Your Dog

Step 1

Use a 1-quart (1-L), wide-mouth, glass Mason jar, this will provide you lots of room for adding water and growing sprouts. Add 1 to 7 tablespoons of sprouting seeds (every 1 tablespoon yields about 1 cup of sprouts).

For the lid, use a cheesecloth over the mouth of the jar and use a rubber band or a mason jar ring to secure it in place. We like dedicated sprouting lids with built-in screens because they make rinsing super easy.

Sanitize Seeds: Fill jar with filtered water, enough to cover seeds plus an inch more. Add a bacteria-killing solution: we use a tablespoon apple cider vinegar plus a drop of liquid dish soap. Let sit for 10 minutes, then rinse **very well** with fresh water, (we rinse up to 7 times).

Step 2

Once seeds are clean, add fresh filtered water to cover the seeds completely, with at least an inch of water above them. Soak for 8 hours or overnight.

Step 3

After 8 hours, drain the water from the jar (we the water we pour off to water our houseplants!). Add fresh filtered water through the lid and swirl the seeds around to rinse them.

Completely drain the water & rest the jar at an angle, so any remaining water can drain out. Rinse and drain the seeds at least twice a day, (ie. morning - night) for 3-5 days.

Step 4

After about a day, the seeds will begin to crack open. Sprouting has begun!

By the 3rd or 4th day, your sprouts will be long enough to eat. Place the jar in a sunny windowsill for a couple hours, and the sprouts will develop a nice green chlorophyll color. Rinse well with the cap off, removing the seed hulls. Drain well, removing all excess water. Store in the refrigerator. Consume within 5 days.

Last Step

Now chop up and add to your dog's food! (start with a teaspoon per 20 pounds of body weight). They can also be frozen and added to your salads or smoothies!

one (of which mushrooms are the highest dietary source), and the substances that promote production of glutathione—selenium and alpha-lipoic acid. Mushrooms also deliver a good dose of polyamines, including spermidine, compounds that increase autophagy and that are found in high levels among centenarians. In animal models, spermidine improves cognition and exerts a neuroprotective effect, likely due to the compound's effects on mitochondrial health. Medicinal mushrooms, including shiitake, maitake, oyster, reishi, lion's mane, turkey tail, Cordyceps, button, and king, are, in fact, the best source of spermidine, a powerhouse longevity molecule. Animals consuming spermidine are also less likely to get liver fibrosis and cancerous liver tumors, even when predisposed. Most impressive, spermidine increases life span, *by a lot.* "It's a dramatic increase . . . as much as 25 percent," says Leyuan Liu, assistant professor at the Texas A&M Institute of Biosciences. "In human terms, that would mean instead of living to about 81 years old, the average American could live to be over 100."

For immune health, 'shrooms save the day with their beta-glucans, a special immune-modulating compound that controls inflammation and keeps insulin low and steady. Recent studies on insulin-resistant obese dogs have revealed the power of beta-glucan supplementation: decreased begging behaviors and decreased appetite. Beta-glucans are found in all edible mushrooms. In addition to helping keep your dog's immune system balanced and reducing inflammation, they exert a positive effect on immunosuppressed dogs and enhance your dog's humoral immune response to vaccines. As for cancer? Individuals who eat 18 grams of mushrooms—or about ⅛ to ¼ cup—daily have a 45% lower risk of cancer compared to those who do not eat mushrooms. For dogs, the median survival time for hemangiosarcoma of the spleen is eighty-six days, but dogs lived *beyond a year* when turkey tail mushroom was added as a sole form of treatment. Medicinal mushrooms are truly remarkable for our health. One of the most obscure mushrooms we use on a regular basis in all realms of life is Chaga, in the form of tea. Chaga is a strange mushroom because its texture is similar to woody tree bark (so there's no sautéing this particular gem). Its hard texture lends itself well to brewing supercharged teas and broths. We add small Chaga chunks to everything that requires a volume of pure water: from bathwater (Dr. Becker) and hummingbird feeders (to reduce bacteria growth, Rodney discovered) to homemade batches of kombucha and soak water for seeds we are sprouting. We've had fresh Chaga tea in the fridge ever since we discovered this brilliant beverage, and its subtle vanilla flavor is delicious iced or hot (for humans) and can be used to supercharge freeze-dried or dehydrated dog food, in place of reconstituting these foods with plain water. Chaga's medicinal properties make it a calming paw rinse for removing road salt in the winter and soothing hot spots in the summer (soak a cotton ball with cooled Chaga tea and apply directly to the sore).

The extraordinary thing about mushrooms is they each have unique medicinal benefits, so you can choose what types to feed depending on the type of support you seek for your dog. For general health and wellness, try porcini, white button, shiitake, turkey tail, maitake, reishi, shimeji, or oyster. Turkey tail and Chaga are potent cancer fighters, and lion's mane is the nootropic mushroom, meaning it nourishes the central nervous system. In addition to glutathione, mushrooms also contain another antioxidant hard to get elsewhere: ergothioneine, aka ergo. Ergo has been named a longevity vitamin by some because, in studies, the molecule has been shown to increase anti-inflammatory hormones and decrease oxidative stress factors in humans. Ergo is found in only one food group: mushrooms. **Chopped medicinal mushrooms are excellent meal toppers**. You can also make Medicinal Mushroom Broth and add it to meals (use it to reconstitute dehydrated or freeze-dried food, plus it makes a great "gravy" over dry food). Mushroom broth ice cubes also make a refreshing treat during the summer months. Dehydrated mushrooms are also fine to use.

LONGEVITY JUNKIE MEDICINAL MUSHROOM BROTH

Add 1 cup fresh (or ½ cup dried) chopped mushrooms and 12 cups pure water (or Longevity Junkie Homemade Bone Broth) to a pot. Grate ½ teaspoon fresh ginger and turmeric root into broth, if desired. Simmer for 20 minutes, and let cool. Puree mixture until smooth, pour into ice cube tray, and freeze. Pop out one portion (1 ounce) for every 10 pounds of body weight, thaw, and mix into food for an instant ergo boost.

All mushrooms that are safe for people to eat are safe for dogs to eat. All mushrooms that are poisonous for people are poisonous for dogs. You can share your cooked or raw mushrooms with your dog as a snack or meal topper. We've found most dogs don't mind mushrooms mixed in with their food, but if your dog won't eat them, they come in supplement form (see Chapter 8). Find a way to get these little miracle workers into your dog's body.

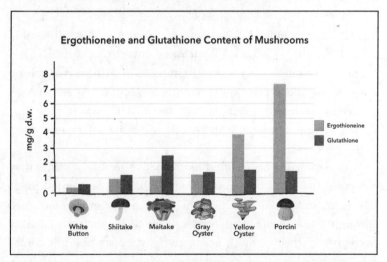

Adapted from Michael D. Kalaras et al., "Mushrooms: A Rich Source of Antioxidants Ergothioneine and Glutathione," *Food Chemistry* (October 2017): 429–33.

Mending the Microbiome

As we described in Part I, the microbial tribes that live in and on our bodies, especially the bacteria that thrive in our and our dogs' guts, are key to health—so much so that it's been said that the gut is like a "second brain." Through a fascinating two-way connection between the gut and brain, the central nervous system tells the brain what's happening in the intestines, and the brain sends a signal back to ensure they're operating smoothly. This exchange of information helps regulate eating, digestion, and sleep patterns. The gut also sends the brain hormonal signals that alert it about hunger, satiety, or pain from intestinal inflammation.

The gut is central to our general well-being: how we feel, our sleep quality, our energy levels, the strength of our immune system, our pain levels, and the efficacy of our digestion and metabolism—even how we think. Researchers are looking at the potential role of some strains of intestinal bacteria in obesity, inflammatory, functional GI disorders, chronic pain, and mood disorders including depression. This research also extends to veterinary medicine. Scientists are finding that nourishing a healthy microbiome by consuming "clean," highly digestible foods (with lower levels of pesticides, contaminants, AGEs, and chemical residues that negatively impact the gut barrier), along with eating more pre- and probiotic foods, can reduce stress-induced diarrhea, combat obesity and inflammation, and support a strong immune system—all of which affect our dog's aging process.

You've probably heard of probiotic foods ("for life") before; they comprise the good bacteria consumed through fermented foods like kefir, sauerkraut, and kimchi. They also can be consumed through supplements. *Pre*biotics are the foods that gut bacteria need to fuel their growth and activity; they are the gut bacteria's preferred meal and are largely made up of indigestible fibers. As with probiotics, they can be ingested through certain prebiotic-rich foods. As the gut bacteria metabolize these otherwise nondigestible, fiber-rich foods, they produce short-chain fatty acids—biomolecules that are beneficial to and even help meet a body's energy needs.

Obviously, we want to support this microbial community and its internal network with the body's biology simply by ingesting the most critical ingredient for gut health: beneficial gut bacteria. One of the star probiotics making headlines in the dog space is *Akkermansia muciniphila* (it's a mouthful, so we abbreviate to *A. muciniphila*). This bacterial species has been shown to promote healthy aging by protecting the mucosal lining of the gut and supporting gastrointestinal health to keep things moving along, while preventing GI conditions like diarrhea and irritable bowel syndrome (IBS). It's also being studied as an agent to combat obesity in pets. *A. muciniphila*'s favorite foods are inulin-rich veggies (like asparagus and dandelion) and bananas. Science says the more *A. muciniphila* bacteria dogs have, the younger they will be. More inulin-rich foods means more *A. muciniphila*, which is a good thing. We recommend feeding prebiotic fibers (like inulin) in the form of whole foods, rather than in supplement form: The sum of the food constituents offers a greater benefit than a single supplement. So many dogs have gut issues, and feeding foods that nourish the microbiome can heal and repair inflamed, dysbiotic intestinal tracts. These microbiome-building foods have *so many more* epigenetic benefits, in addition to fostering a healthy gut.

MICROBIOME-BUILDING CORE TOPPERS
YOU CAN SHARE WITH YOUR DOG

➤ **Endive, escarole, and radicchio:** All members of the chicory family are fine to use as toppers to any base food. These greens are loaded with prebiotic fiber that feeds the beneficial bugs in your dog's gut.

➤ **Dandelion:** You and your dog can eat all parts of the
dandelion—the flowers, stems, leaves, and roots. Dandelions
are loaded with prebiotic fiber and have a cleansing effect on
the liver and bloodstream. Dandelions are more nutritious
than kale and are chock-full of vitamins (C, beta-carotene, and
K) and potassium. Now that's a free medicine cabinet in your
backyard! (Just make sure they have not been sprayed.) You also
can find fresh dandelion greens in many grocery stores.

➤ **Okra and asparagus** are great sources of not just prebiotics
and vitamins. Asparagus is one of the few foods that naturally
contains glutathione, a fundamental chemical serving as a
master internal antioxidant and detoxifier that the brain loves.
Both of these veggies can be sliced raw for perfect training
treats or can be steamed and shared in the bowl.

➤ **Cruciferous veggies like broccoli and arugula:** In addition to
being loaded with gut-friendly fiber, cruciferous vegetables
also come with vitamins, antioxidants, and substances with
anti-inflammatory effects. Broccoli in particular contains
two supermolecules: 3,3'-Diindolylmethane (DIM) and
sulforaphane, which naturally boost glutathione levels. DIM
helps the body manage a healthy balance of hormones and
clear xenoestrogens (environmental chemicals that mimic
estrogen) that can muck up the system. Research in dogs also
shows DIM can also have antitumor/anticancer activity. And
sulforaphane has been studied for bone and bladder cancer
in dogs with impressive results. A key point: The magic
of sulforaphane is found only when broccoli is eaten; dogs
and people don't benefit from sulforaphane in supplement
form because it degrades so quickly. This magical molecule
stimulates apoptosis (healthy, programmed cell death) in your
dog's body, which is vital when bad cancerous cells need to be
killed. Small broccoli pieces and chopped stems make great
training treats; alternatively, recycle the cooked broccoli your

family had for dinner (without sauces) into your dog's bowl. If your dog has never had broccoli or brussels sprouts, gentle steaming can reduce gas production until your dog's body adjusts to the new veggies.

Do cruciferous veggies cause low thyroid? You may have heard that a very high intake of cruciferous vegetables (much more than dogs would naturally eat) has been linked to hypothyroidism (having low thyroid hormones). In lab rodents, this was found to be a result of a metabolite in the vegetables—thiocyanate—competing with iodine (a mineral required for thyroid hormone production) for uptake by the thyroid gland. Thankfully, additional animal research conclusively demonstrates that increased cruciferous vegetable consumption does *not* appear to increase the risk of hypothyroidism *unless* it's accompanied by iodine deficiency. As long as you are feeding a nutritionally complete diet, there's no reason to fear this family of veggies!

➤ **Jicama:** This crispy vegetable tastes like a cross between an apple and a potato and makes for a perfect training treat. Jicama is incredibly high in the prebiotic fiber inulin as well as vitamin C.

➤ **Jerusalem artichoke (sunroot, sunchoke, or earth apple):** This knotty tuber root vegetable, which is not related to globe artichokes, comes from the sunflower family and is packed with inulin. Some nutritionists believe sunchokes are the unsung heroes in the veggie root family because they're so versatile and deliver a punch of prebiotics.

➤ **Fermented veggies,** either store-bought or homemade, are a rich source of probiotics for dogs. The problem is getting dogs to eat these twangy, tart blends. If your dog will indulge, make sure they're onion-free and give one quarter of a teaspoon for every ten pounds once a day, mixed in food.

Forever Dog Fruits

Avocados: This bumpy on the outside, creamy on the inside green fruit boasts significant amounts of vitamins C and E and potassium, and a lot of folate and fiber. Avocados also contain oleic acid, a healthy monounsaturated fat also found in olive oil. Oleic acid helps with brain function and boosts our health at every age. Emerging research shows that avocados also benefit skin, eye, and even joint health. Avocados also contain heart-friendly phytosterols such as beta-sitosterol.

Green bananas: Bananas provide potassium, but they're also high in sugar when fully ripe (one medium banana contains fourteen grams of sugar—that's three and a half teaspoons!). Unripe tropical fruit, on the other hand, has lower fructose content and is made up of resistant starches, which feed your dog's microbiome. In addition, they offer antioxidant, anticancer, and anti-inflammatory tannins, along with carotenoids that help prevent oxidative stress. So seek out the greenest bananas and cube them up the size of peas for tiny training treats.

Raspberries, blackberries, mulberries, blueberries: Berries are full of prebiotic fiber and polyphenols like ellagic acid. Kriya Dunlap, PhD, and her colleagues at the University of Alaska, Fairbanks, found that diets supplemented with fruits high in antioxidant compounds can potentially serve to sustain the body's antioxidant levels and prevent exercise-induced oxidative damage. Her research focused on sled dogs, who are constantly vulnerable to muscle damage associated with the rigors of their exercise. In blueberry-fed dogs, she showed that they enjoyed a greater total amount of antioxidants in their blood plasma immediately postexercise, which better protected them against the deleterious effects of oxidative stress. We use a lot of frozen blueberries as training treats, when fresh blueberries aren't in season. But be forewarned: more than one blueberry for every two pounds of body weight a day (i.e., five blueberries for a ten-pound dog) can lead to completely benign dark blue poop, so feed a few, then move on to other Longevity Foods as healthy rewards that day.

Strawberries: These red jewels get an extra special shout-out because they contain a little-known antiaging secret called "fisetin," a plant compound that researchers have long studied for its antioxidant and anti-inflammatory properties. Recently, scientists have discovered it also kills senescent cells—those zombie cells that are a hallmark of premature aging. A cell study published in the journal *Aging* showed that it eliminated about **70 percent** of senescent cells—while doing no harm to healthy, normal cells. Cell senescence can be defined as cells that cannot divide but aren't dead, so they accumulate and cause inflammation in other cells. In one striking study, mice exposed to fisetin lived 10 percent longer and experienced less age-related issues than the control group, even at an older age. These findings prompted the Mayo Clinic to sponsor an ongoing clinical trial examining the direct effects of fisetin supplementation on age-related dysfunction in humans. Fisetin also mimics all the positive effects of fasting (including reducing mTOR and boosting AMPK and autophagy) in addition to protecting the heart and nervous system. Occasionally, you'll read information that suggests avoiding feeding strawberries to dogs; this comes from the rare possibility that dogs can consume too many *green leafy strawberry stems* and can cause an upset stomach. Removing the green stem removes any risk of GI upset. Choose spray-free or organic strawberries.

Pomegranate: Pomegranate has been shown to help protect cells and, especially, the heart. Cardiac diseases are considered to be the second-most-prevalent cause of death for dogs. Valvular endocardiosis and dilated cardiomyopathy are the most common of these and are known to be more prevalent in aged dogs. Oxidative damage leading to cell death may constitute one of the most proximal of the cascade of events that results in heart failure. In a study published in the *Journal of Applied Research in Veterinary Medicine*, scientists found that feeding pomegranate extract to a dog has incredible protective heart and health benefits. Pomegranates also contain molecules called "elligantans," which our gut microbes convert to urolithin A. Urolithin A has been shown to regenerate mitochondria in worms—increasing their life span by more than 45 percent. These results motivated scientists to replicate these tests in rodents. They met with similar results. Older mice showed signs of increased mitophagy (self-destruction of damaged mitochondria) and demonstrated better running endurance compared to the control group. You'd be surprised how many dogs will eat these tart, crunchy little jewels mixed in with their food, about a teaspoon for every ten pounds. If your dog isn't one of them, keep reading; you'll find something they love.

POWERFUL PROTEINS

Sardines: Did you know that sardines are named after Sardinia, the Italian island where large schools of these fish were once found—and where people tend to live long, healthy lives? It's a Blue Zone, where a disproportionate number of people live past one hundred compared to the rest of the world. Sardines may be small fish, but they are big in nutrients, rich in omega-3 fatty acids, vitamin D, and vitamin B12—key players in the longevity game. Buy sardines packed in water (or fresh, if you can get 'em). A sardine for every twenty pounds of body weight, two or three times a week, will do the trick.

Eggs: Whether from chickens, quails, or ducks, eggs are nature's nutrition bombs, packed with vitamins, minerals, protein, and healthy fat. Eggs are also rich in choline, a nutrient crucial to the production of the neurotransmitter acetylcholine in the brain, which aids brain function and memory. Eggs can serve a dog's body in various ways, because the amino acid profile of egg protein matches dogs' biological needs. Raw, poached, hard-boiled, scrambled—it doesn't matter, dogs like them most ways. Go for pastured, free-range eggs to get the most nutrients. One egg is about seventy calories. Thirty-pound Shubie gets an egg on her meal several times a week.

Organ meats: Liver, kidney, tripe, tongue, spleen, pancreas, heart...many of us can't fathom the attraction, but dogs love them all. Free-range organ meats are alpha-lipoic-acid-rich delicacies that make for great training treats fed raw, freeze-dried, dehydrated, or cooked and cubed. Dogs will always try to score more organ meats from you as treats, but they're rich so use the "paw principle" to meter daily intake: The size of *one* of your dog's paws, width plus length (and a depth of where the pad turns to hair), is a reasonable amount of daily organ meat treats for most healthy, active dogs. The smaller you make the pieces, the more pieces she gets to eat! Other healthy proteins to share with your dogs are sardines, cod, Alaskan halibut, herring, freshwater fish, chicken, turkey, emu, pheasant, quail, lamb, beef, bison, elk, venison, rabbit, goat, kangaroo, alligator (if your dog has allergies to the other options), and cooked wild salmon. Any and all lean, clean, uncured meats make great training treats for dogs. Meats *not* to share with your dog: cured meats, ham, bacon, pickled herring, smoked meats, sausages, and *raw* salmon.

Grab-and-Go Treats:
The Doggie Dictionary of Fresh Pharma

ANTIOXIDANT RICH	
vitamin C filled	bell peppers
capsanthin filled	red bell peppers
anthocyanin rich	blueberries, blackberries, and raspberries
beta-carotene rich	cantaloupe
naringenin filled	cherry tomatoes

ANTIOXIDANT RICH (CONT.)	
punicalagin loaded	pomegranate seeds
polyacetylene loaded	carrots
apigenin loaded	peas
sulforaphane rich	broccoli

ANTI-INFLAMMATORY	
bromelain filled	pineapple
omega-3 dense	sardines (except for dogs requiring a low-purine diet)
quercetin rich	cranberries (not for finicky dogs)
cucurbitacin rich	cucumbers

SUPERFOOD CALIBER	
choline filled	hard-boiled eggs
glutathione filled	button mushrooms
manganese rich	coconut meat (or dried, unsweetened coconut chips)
vitamin E rich	raw sunflower seeds (sprout them and other microgreens for a chlorophyll-rich upgrade from grass!)
magnesium filled	raw pumpkin seeds (feed all seeds one at a time for perfectly sized training treats, up to $1/4$ teaspoon for every ten pounds of body weight, spread throughout the day)
selenium rich	Brazil nuts (chop up one a day for you and your big dog, or share one with smaller dogs)
folate filled	green beans

SUPERFOOD CALIBER (CONT.)	
fisetin filled	strawberries
indole-3-carbinol rich	kale (or homemade kale chips)
isothiocyanate loaded	cauliflower

DETOX DELIGHTS	
apigenin loaded	celery
anethole filled	fennel
fucoidin rich	nori (and other seaweeds)
betaine filled	beetroot (except for dogs with oxalate issues)

FOR GUTS AND GLORY	
prebiotic rich	jicama, green bananas, sunchokes, asparagus, pumpkin (great filling for food-enrichment toys and food puzzles)
actinidin rich	kiwi
pectin rich	apples
papain rich	papaya

Herbs for Health Span

Herbs and spices have a long, rich history in many cultures around the world not only for flavoring foods but for healing the body and preventing disease. Some plants have developed a richer, more diverse spectrum of bioactive phytochemicals that, even when consumed in tiny quantities, exert a profoundly positive effect on a variety of organ systems and biochemical pathways. Using medicinal herbs (many found right in your kitchen spice drawer or garden) is an easy and inexpensive way to provide powerful plant compounds directly to your dog's dish.

If it's been a while since you've looked at the expiration date of your culinary spices, we recommend starting with fresh bottles, preferably organic. **One shake of dried herbs for every ten pounds of dog** is a good rule when "spicing up" your dog's bowl. The worst thing that can happen if you over-season your dog's meals with healthy herbs is you discover you love cilantro more than your dog does, so start with a light sprinkling of culinary herbs until you know your dog's preferences. Mix herbs into food prior to feeding. Fresh herbs can be added, finely minced, at one-quarter teaspoon per twenty pounds of body weight per day. Dried herbs are more potent than fresh herbs, but we've found dogs generally readily accept both forms mixed into their food.

Parsley: No longer a garnish you throw out, this herb (an apiaceous vegetable) has lots of reasons to boast. One bioactive compound neutralizes carcinogens and prevents oxidative damage by activating glutathione-S-transferase (GST), which stimulates glutathione production (needed to clear AGEs from the body). In animal models, parsley's volatile oils can raise the free radical–quenching capacity of the blood and help neutralize carcinogens, including the benzopyrenes produced from high-heat-processed foods.

Turmeric: The medical literature exploring the efficacy of curcumin, the most active polyphenol in the Indian spice turmeric (pronounced "too"-meric), continues to explode, with thousands of published studies available today. The active ingredient in turmeric, curcumin, has been shown to help boost the levels of brain-derived neurotrophic factor (BDNF) and improve cognition. A 2015 study showed its neuroprotective effects in dogs by targeting biochemical pathways associated with neurodegenerative disorders, including cognitive impairments, energy/fatigue, mood, and anxiety.

Turmeric is truly a jack-of-all-trades; its uses could fill an entire reference manual. In 2020, for example, researchers at Texas A&M showed turmeric's promise in decreasing ocular inflammation in dogs suffering from uveitis, an inflammation of the eye that leads to pain and reduced vision. We have both used this amazing root to manage inflammation from a variety of causes from head to tail; it's among our most favorite add-ins. Ethnobotanist James Duke published a meta-study of over seven hundred turmeric studies and concluded: "Turmeric appears to outperform many pharmaceuticals in its effects against several chronic debilitating diseases, and does so with virtually no adverse side effects." When turmeric is combined with rosemary, it has a synergistic effect on canine mammary carcinoma (breast cancer), mastocytoma, and osteosarcoma cell lines, and this combination had an additive effect with chemotherapeutic agents.

Rosemary: This herb is being investigated as the "spice of life" because it contains 1,8-cineole, a compound that boosts acetylcholine production in the brain and reduces cognitive decline. It also boosts your dog's BDNF levels. Rosemary's antioxidant and anti-inflammatory qualities are largely attributed to its polyphenolic compounds, including rosmarinic acid and carnosic acid, both of which have anticancer effects. Moreover, carnosic acid may aid in promoting eye health by preventing cataracts, common in both dogs and humans.

Cilantro: Cilantro (coriander) is a power-packed herbal gem that contains a generous amount of antioxidants in the form of phytonutrients. It also contains active phenolic compounds, manganese and magnesium. No wonder cilantro has been used as a digestive aid, an anti-inflammatory, and an antibacterial agent; and as a weapon in the fight to control blood sugar levels, cholesterol, and free-radical production. There's also science showing how cilantro can naturally help eliminate lead and mercury from the body through the urine, one of the reasons we recommend using it periodically for detoxification purposes.

Cumin: *Cuminum cyminum* has many health benefits: It improves digestion and it's antifungal, antibacterial, and potentially anticancer.

Cinnamon: Culled from the rolled bark of a South Asian tree and one of the most beloved super spices, cinnamon is most famous for its collagen-building assets. Collagen is one of the body's most abundant (and important) proteins and is extra important for aging dog joints. Cinnamon also is gaining more fame for its aid in balancing blood sugar and its antioxidants that help protect the cardiovascular system by managing oxidative stress, reducing inflammatory reactions, and reducing circulating fats. Cinnamaldehyde, an active ingredient in cinnamon, is being studied in animals for its ability to prevent neurodegenerative diseases, including Alzheimer's. In one clinical study, cinnamon improved all tested heart parameters in dogs after only two weeks. If you add a dash of cinnamon to your dog's food, make sure to mix it in so your dog doesn't inhale the fine powder.

Cloves: Apart from being a rich source of manganese (an important and hard-to-come-by mineral needed to keep your dog's tendons and ligaments in good working order), cloves contain the antioxidant eugenol, which prevents oxidative damage caused by free radicals five times more effectively than vitamin E. Eugenol may in fact be especially beneficial for the liver. One animal study fed rats with fatty liver disease mixtures containing either clove oil or eugenol. Both mixtures improved liver function, reduced inflammation, and decreased oxidative stress. Cloves have free radical–scavenging properties and contain antioxidants that slow down signs of aging and reduce inflammation. Other studies have looked into cloves' anticancer and antibacterial properties with promising results. Whole cloves can be a choking risk, so grind them prior to feeding and add a tiny pinch for every twenty pounds of body weight.

OTHER HEALTH SPAN HERBS YOU CAN SHARE FROM YOUR SPICE CABINET OR GARDEN

Basil: In addition to supporting heart health, basil helps manage the body's stress load by lowering levels of cortisol.

Oregano: This one is big in antibacterial, antifungal, and antioxidant action, with loads of vitamin K to boot!

Thyme: Thyme contains thymol and carvacrol, which have potent antimicrobial properties.

Ginger: Well known as the anti-nausea herb, gingerols also delay aging by addressing oxidative stress in animals and serve a neuroprotective purpose.

Hazard Alert: What culinary herbs are no-nos for dogs? Don't feed your dog chives (a member of the onion family) or nutmeg (a rich source of myristicin, which can cause neurologic and GI symptoms if ingested in moderate quantities).

Forever Fluids

For thousands of years human beings have supercharged their nutritional intake by consuming plant extracts, juices, or infusions of plants and herbs. Most of us never consider adding concentrated medicinal infusions into our meals or drinks. But when it comes to feeding our dogs, medicinal teas can add longevity benefits to every meal at a low price. In particular, cooled teas are an economical and powerful way to deliver a plant's best medicinal properties directly to your dog. Herbal teas are naturally decaffeinated. Green and black teas should be decaffeinated. Buy organic teas, if possible.

All teas can be steeped as usual (we recommend starting with one teabag in three cups of very hot, purified water), then cooled off prior to adding to your dog's food. Alternatively, add warm tea to your dog's dry food and allow the therapeutic brew to marinate with the kibble, creating a super-saucy gravy and adding hydration (dogs aren't designed to eat low-moisture, dried food their whole lives; teas help!). If you feed dehydrated or freeze-dried dog food, reconstitute them prior to feeding with tea or our Longevity Junkie Homemade Bone Broth (page 242). It's fine to blend different teas together or use a specific tea for a specific purpose. Below are some highlights from the science.

Decaf green tea: If you're health conscious, then you already know that green tea is good for you. It's been written about extensively in both medical and lay literature because its healthy bioactive compounds have powerful anti-inflammatory, antioxidant, and pro-immune effects. Its properties have long been documented to improve brain function, protect against cancer, lower risk of heart disease, and enhance fat loss. Multiple studies have arrived at the same conclusion: People who drink green tea may live longer than those who don't. So it should come as no surprise that green tea extract has been used in pet food for quite some time and as a therapeutic agent for obesity, liver inflammation, antioxidant support, and even radiation exposure in dogs.

Decaf black tea: Like green tea, black tea is rich in polyphenols, the natural chemicals that act as strong antioxidants that help remove free radicals from cells and tissues. Polyphenols are partly what make these teas anticancer and anti-inflammatory. The difference between green and black tea is that black tea is oxidized and green tea is not. To make black tea, the leaves are first rolled and then exposed to air to trigger the oxidation process. This reaction causes the leaves to turn dark brown and allows the flavors to heighten and intensify. The type and amount of polyphenols in black and green teas differ. For example, green tea contains a much higher amount of epigallocatechin-3-gallate (EGCG), which is technically a catechin that helps limit free-radical damage and protects against cellular damage. Black tea is a rich source of theaflavins, a group of antioxidant molecules that are created from catechins. Both types of tea have similar effects on protecting the heart and boosting brain function. And both contain the calming amino acid L-theanine, which alleviates stress and calms the body.

Mushroom teas: All are healthy and safe to offer dogs. The two teas dogs tend to like the most are:

- ➤ **Chaga tea:** As noted earlier, Chaga is a type of medicinal mushroom from which a tea can be derived. Packed with antioxidants, Chaga mushroom's extract may fight cancer and improve immunity, chronic inflammation, blood sugar, and cholesterol levels. More studies on this tea are underway, especially with regard to its implications for learning and memory.

- ➤ **Reishi tea:** Another tea made from the reishi mushroom (used for centuries in Eastern medicine), this tea owes its health benefits to several molecules, including triterpenoids, polysaccharides, and peptidoglycans, that can boost the immune system, fight cancer, and improve mood.

Calming teas: Fear, anxiety, and restlessness are some of the most common stress behaviors in dogs, and there are lots of herbal teas that may help: chamomile, valerian, lavender, and holy basil (tulsi). All can be brewed, cooled, and added to your dog's food prior to serving.

Detox teas: In the detox tea department, we have dandelion, burdock, and oregano leaf. We won't get into the details of these teas' bountiful pro-health benefits except to say you can't go wrong enjoying a tea party with your furry companion. And you needn't go far. In fact, there are many other types of common plants you may have in your garden that can be made into wonderful teas for your dog, including rose hip, peppermint, lemon verbena, lemon balm, lemongrass, linden flower, calendula, basil, and fennel.

Tip: You can add herbal tea bags to bone broth to create a synergistic solution of high-powered micronutrients. Pour into ice cube trays and freeze, then pop out one for every ten pounds of body weight a day.

LONGEVITY JUNKIE HOMEMADE BONE BROTH

This bone broth recipe is different than traditional recipes, which can be high in histamines that can negatively affect some dogs.

Cover a free-range, organic, whole chicken (or leftover carcass or raw soup bones of your choice) with pure filtered water, and add:

- ½ cup chopped fresh cilantro (effective at binding heavy metals)

- ½ cup chopped fresh parsley (a natural blood detoxifier)

- ½ cup chopped fresh medicinal mushrooms (providing glutathione, spermidine, ergo, and beta-glucans)

- ½ cup cruciferous veggies, such as broccoli, cabbage, or brussels sprouts (these foods have the high sulfur content needed for liver detoxification)

- 4 cloves raw garlic, chopped (the high sulfur content stimulates production of glutathione for liver detoxification)

- 1 tablespoon unfiltered, raw apple cider vinegar

- 1 teaspoon Himalayan salt

Cover and simmer 4 hours. Turn off stove. Add 4 tea bags, if desired. Steep tea in broth for 10 minutes, then discard tea bags. Remove any remaining meat from bones and discard bones. Puree remaining meat, veggies, and liquid until a smooth, gravy consistency. Freeze in individual portions (ice cube trays work well). Remove portion (most standard trays are one ounce/portion, or two tablespoons; use a cube for every ten pounds of body weight) and let thaw to room temperature or reheat to warm the broth prior to adding it to your dog's food.

Herbaceous Health Hoaxes: How Did We End Up Fearing So Many Foods?

We're floored at the misinformation we've found on the Internet surrounding what you can and can't feed to dogs. What foods are genuinely toxic to dogs? The European pet food industry federation (fediaf.org) publishes the most accurate, science-backed information regarding food toxicity in pets. Significantly, it lists only *three* foods as being toxic to dogs and cats: grapes (and raisins), cocoa (chocolate), and members of the onion family (including onions, chives, and high doses of *garlic extract* (meaning garlic supplements; fresh garlic is fine).

Compare Europe's short no-no list (three foods and one supplement) with the extensive lists offered by ASPCA, AKC, and dozens of other online resources purporting to identify "foods that are toxic to pets." The comparison will make your head spin. The vast majority of online no-no lists include foods that are *genuinely toxic to dogs* (FEDIAF's list of three foods and one supplement), foods that should be avoided in pets with specific medical issues, and foods that can pose a choking hazard. For instance, dogs with pancreatitis (inflammation of the pancreas) should avoid all cooked fats and high-fat foods while they are recovering from this medical condition. Many websites list eggs, seeds, and nuts as "toxic" because these foods are higher in healthy fats and can exacerbate pancreatitis. But eggs, seeds, and nuts (excluding macadamias, which don't contain an identifiable toxin but do cause nausea from their very high fat content) are in and of themselves not toxic to dogs; they are healthy and nutritious foods that can and should be fed to healthy dogs. Similarly, a host of nutritious foods, including raw almonds, peaches, tomatoes, cherries, and a wad of other very healthy fruits and vegetables are listed as "toxic" because they can pose a choking hazard if the pits aren't removed or the animal eats the whole plant, not just the fruit.

Unfortunately, **the truly systemically toxic foods (all four of them) have been lumped together with the "not appropriate for all medical conditions" and choking risk foods to create a massive list of no-no foods that dog owners downright fear.** And for no good reason. Common sense (like removing the pit prior to offering your dog a piece of apricot) and cited research (e.g., toxicity studies) support a very different approach to canine nutrition. Do your own research and you'll likely end up where we did (after an extensive literature review): **Don't feed grapes (or raisins), onions, chocolate, or macadamia nuts to any dog, ever. That's it. Otherwise, use common sense.** European common sense for the win.

Here are some **canine urban food myths** we can put to rest once and for all:

➤ "Avocados and garlic are toxic." FALSE. Don't feed or eat the skin or pit of avocados, as they contain a substance called "persin" that can cause GI upset, but the flesh is safe and healthy for you and your dog. We smear a wedge the size of an orange slice (approximately 40 calories) into Shubie's Kong toy once a day. See note about garlic, below.

➤ "Never feed dogs mushrooms." FALSE. Mushrooms that are safe for people are safe for dogs. Mushrooms that are highly medicinal for humans are highly medicinal for dogs (and the same goes for toxicity). A tablespoon per twenty-five pounds of body weight is a nice place to start!

➤ "Rosemary causes seizures." SOMEONE'S CONFUSED. The *essential oils* of rosemary and eucalyptus (the volatile, potent aromatic oils you can buy at health food stores) contain a concentrated amount of camphor, a compound that, *if consumed by epileptics*, can increase seizure potential. (We agree: don't feed your epileptic dogs large quantities of rosemary essential oil.) A tiny pinch of fresh rosemary or a dash of dried rosemary and other herbs added to your healthy dog's bowl are *minute* quantities, enough to stimulate a positive health outcome but not enough to negatively affect even the most sensitive dog.

➤ "Walnuts are toxic." PSEUDOSCIENCE. Raw, unsalted English walnuts (and almonds and Brazil nuts) certainly can be a choking hazard for dogs, so break them up into small pieces prior to feeding. One walnut half can be chopped into four perfect training treats for a fifty-pound dog and offered throughout the day. Again, the only nuts that are a risk for dogs are macadamia nuts, which can cause nausea. Peanuts may contain some mycotoxins but they are not innately toxic to dogs. If you have a black walnut tree in your yard, don't let your dog eat the bark (it can cause neurological symptoms) or the thick husk that encases the hard nut, as the mycotoxins that sometimes grow on the outer skin can cause vomiting.

Note about garlic: Garlic gets a bad rap in vet medicine because it's in the onion family. Onions contain about fifteen times the thiosulfate concentration of garlic, the compound responsible for causing Heinz body anemia when dogs eat onions. A 2004 study demonstrates that allicin, the medicinal compound in garlic, is beneficial for animal cardiovascular health as well, with no report of anemia, despite high concentrations fed during the study (this is the reason you see garlic included in lots of commercial pet foods and vets being fine with it). Below are the recommended doses of *fresh* garlic per day, based on weight, if you choose to feed this medicinal spice (we don't recommend garlic pills):

➤ Ten to fifteen pounds—½ clove
➤ Twenty to forty pounds—1 clove
➤ Forty-five to seventy pounds—1½ cloves
➤ Seventy-five to ninety pounds—2 cloves
➤ One hundred pounds and over—2½ cloves

LONGEVITY JUNKIE TAKEAWAY

➤ The 10 Percent Rule: Ten percent of your dog's calories can come from healthy human food "treats" and not "rock the balance."

➤ You don't have to dramatically overhaul your dog's diet overnight. Start small and easy, swapping out carby, highly processed snacks for proven Longevity Foods like dog-friendly fresh fruits and veggies. Or add a dollop of them to your dog's bowl with whatever you've been serving him until now. Recycle the dinged, dented, and trimmed pieces of veggies into your dog's bowl that you'd otherwise toss.

➤ Examples of easy and convenient Longevity Treats: raw, chopped carrots; pieces of apple; broccoli; cucumber; berries; apricots; pears; peas; pineapple; plums; peaches; parsnips; cherry tomatoes; celery; coconut; pomegranate seeds; raw pumpkin seeds; mushrooms; hard-boiled eggs; zucchini; brussels sprouts; and cubed meats or organs.

➤ A great way to naturally support your dog's microbiome is to offer prebiotic-rich veggies like asparagus, green bananas, okra, broccoli, Jerusalem artichokes (sunchokes), and dandelion greens.

➤ Teas, spices, and herbs are excellent sources of longevity medicine for your dog.

➤ Two homemade recipes to try are Longevity Junkie Medicinal Mushroom Broth (page 225) and Longevity Junkie Homemade Bone Broth (page 242).

➤ Contrary to many urban myths, not many foods are truly toxic to dogs. Grapes (and raisins), onions (and chives), chocolate, and macadamias are a definite no. Avoid nutmeg, too.

8

Supplemental Habits for a Long and Healthy Life

The Essentials to Navigate Safe and Effective Supplements

> Health is like money, we never have a true
> idea of its value until we lose it.
>
> —Josh Billings

In 2011, Pusuke, a Shiba Inu mix who was recognized the previous year by Guinness World Records as the planet's oldest dog, died at home in Japan at age twenty-six. His owner attributed Pusuke's longevity to twice daily vitamins in addition to lots of love and exercise. We may never know how much the vitamins contributed to Pusuke's good long life (or the exact blend of "vitamins"), but plenty of other anecdotal stories echo the same experience. The good news is that science is finally catching up to all this anecdotal evidence showing that supplements are powerful tools when used properly. In the past decade, more and more studies have flooded the collective library of canine data documenting the value of certain supplements preventing and treating disease or injury, and lengthening life. We are here to separate the wheat from the chaff for you. There are a lot of excellent supplement formulas made today by people who love dogs as much as you do, and who are committed to doing whatever it takes to share these longevity jewels. We should add that many of these studies involving dogs have further informed human health.

Strolling down any aisle in a store that sells supplements can be dizzying if you're not sure what you're looking for and feel overwhelmed by the sheer number of formulas, brands, and health claims screaming out at you. You'll come across names of things you've never heard of before, names you can't pronounce (ashwagandha? phosphatidylserine?). At the same time, you read tantalizing assertions like "Add X and watch your dog thrive," or "Clinically [or "scientifically"] proven to do X, Y, and Z," or the ultimate bait: "Extend the life of your dog by 30 percent or more by taking X!"

The supplement industry is colossal, and colossally confusing, too. But with the right knowledge and a list of trusted recommendations, it can be magical. The pet supplement business has exploded and is on track to be a billion-dollar industry—we're talking just *supplements* here, for the pet food industry alone is nearing $135 billion. The global pet supplements market size was valued at $637.6 million in 2019 and is expected to grow at a compound annual growth rate of 6.4 percent from 2020 to 2027.

The force propelling this market? The same customers who spurred the general wellness movement and self-care culture in the last decade: boomers and millennials. In fact, millennials, for whom pets can be surrogates for kids, have swiftly overtaken their elders as the main drivers of demand for high-quality supplements. It's estimated that between 57 and 65 percent of US households own a pet. The American Pet Productions Association trade group reports the higher percentage, which would mark an all-time high for pet ownership. Currently, the age group that owns the most pets is millennials, and it's safe to assume that they care for them in the way they'd nurture a child. That's because as of 2018, the US birth rate was the lowest it had been in thirty-two years.

TD Ameritrade polled 1,139 millennial pet owners in 2018 and found that 70 percent would take a break from work to care for a new pet. Eighty percent of women and 60 percent of the surveyed men said they believed a pet to be a "fur baby." Pet insurance has soared by 18 percent (from 1.8 million in 2017 to over 2 million in 2018). And the US Bureau of Labor Statistics forecasts that both vet and vet tech jobs will increase by 20 percent by 2028.

For millennials, products that earlier generations considered luxuries are now considered essentials. Even venture capitalists and corporate buyers cannot avoid partaking in the gold rush, as they host summits to court start-up companies committed to developing longevity products and supplements. (Dog owners aged twenty-five to thirty-four skew particularly high for buying supplements for their animals. Overall, dog people spend four times as much on their furry friends as cat people and account for an estimated 78 percent of all pet supplement sales.)

Although the general purpose of supplements is to help fill in gaps that diet does not or cannot adequately supply, sometimes people take supplementation to extremes, which can work against a body. Indeed, there can be too much of a good thing. Antioxidants are a great example. While they are key to controlling free radicals, too many synthetic antioxidants consumed through supplements can hamper the body's own innate antioxidant and detoxification machinery. Our DNA activates the production of internally produced (endogenous) protective antioxidants triggered by certain messages. This entirely natural antioxidant system is more seamless and potent than any supplement.

In periods of high oxidative stress, humans and their canine companions have developed biochemical methods to increase antioxidant production. In turn, these antioxidants protect the cells. We don't rely entirely on food for antioxidants; our bodies possess the means to meet our cells' needs.

Several natural compounds that turn on antioxidant and detoxification pathways have been identified. These pathways often involve a special protein called "Nrf2," which we defined in Chapter 7. Some scientists call this protein a "master regulator" of aging because it activates many genes related to longevity and quells oxidative stress. Among these natural, Nrf2-triggering compounds are curcumin from turmeric, green tea extract, silymarin (milk thistle), bacopa extract, docosahexaenoic acid (DHA), sulforaphane (contained in broccoli, not a supplement), and ashwagandha. Every one of these triggering compounds works to ignite the body's natural production of important antioxidants, including glutathione, one of the most important detoxification agents. In veterinary medicine, studies show that dogs who are not aging well or have naturally occurring liver disease also have low glutathione levels. Glutathione also assists wth detoxification by attaching to some toxins and decreasing their toxicity. It will be among the supplements we recommend you consider, along with the others listed above, some of which fuel a body's own production of glutathione.

FOOD SYNERGISM

The Whole Is More Than the Sum of the Parts

Supplements are not magic bullets or insurance policies against a bad diet; contrary to savvy marketing ploys, they also aren't the secret to immortality. In fact, we don't recommend adding supplements until you've corrected your dog's diet. Supplements are not shortcuts to optimal health. But sometimes supplements are the only way to get the quantity of active substances at the volume needed to shift health. For example, it's not practical for a dog to eat a truckload of apples or kale to gain the benefits of quercetin when a supplement can supply the flavonoid in concentrated, therapeutic amounts. But do try to get most of what's needed from food first. Labradors tend to eat *anything*, Chihuahuas don't. If you need a supplement, give it. Not all dogs need supplements all the time.

We could write an encyclopedia on all of the individual supplements that can help to support specific breeds, medical conditions, and life stages, but many other people have done that and there's a lot of credible information to mine online, too. What's currently not available is a straightforward list of synergistic antiaging/longevity-promoting supplements, so we've done the work for you. We've curated a list of a few we really like in each category, but, of course, there are dozens of others that can be exceptional; check out www.foreverdog.com for a deeper discussion on supplements.

The list is broken down into the core essentials every guardian should consider and the optional add-ons for special needs based on your dog's unique situation (i.e., age, breed, health status, exposures). We recommend that you evaluate all the core basics as they relate to your own dog's lifestyle, and then add other supplements as you see fit given your dog's unique physical needs (also see Chapter 9). Your budget can also help lead the way; for some people, buying lots of extra supplements (or remembering to administer them) is not in the cards, and that's okay. We'll give you the information you need to navigate this tricky area; you get to choose your dog's protocol.

We keep an ongoing and ever-evolving list at www.foreverdog.com as well. Because the supplement industry is not regulated (like FDA-approved drugs), there are quality disparities among brands, and not all pet supplements are made with human-grade raw materials. Companies change hands and products can be discontinued. This can be a dynamic area in which, one day, a large study changes the image of a certain supplement or a new one worth considering comes on the market. The basics that follow are not likely to change anytime soon. If your dog has any diagnosed medical conditions, takes medication, or is scheduled for surgery, consult your vet before starting a new supplement routine.

THE POWER OF THE PULSE

Why We Recommend "Pulse Therapy" for Most Supplements

Giving the same supplements every day means the body has plenty of time to adapt and adjust to the same molecules entering the body, over and over. Switching up the brand and frequency of dosing optimizes the body's response to supplements. For this reason, we recommend some supplements several times a week. It's also okay to forget a day or skip a day, so don't panic. The only pills your dog needs at the same time every day are the prescriptions he is on to manage medical conditions. Supplements don't need to be tied to a rigid schedule. You're providing them to whisper longevity instructions to your dog's epigenome—the second layer of chemical compounds that surrounds your dog's DNA and acts like a cheat sheet for their DNA, modifying gene expression by turning genes off or on.

Core Basics

We talked a lot about the enzyme AMPK, mTOR, and autophagy in Chapter 3, as these things relate to certain cellular housecleaning activities and longevity. Ideally, we want to support the pathways that turn the volume down on mTOR so autophagy can work its magic in the body. As a quick reminder, mTOR is basically the body's biological "dimmer switch" for turning autophagy in cells on or off, which is how cells tidy their rooms and recycle where possible. We also want to spark those antiaging sirtuin genes and the actions of AMPK, the body's antiaging molecule that manages vital cellular housecleaning and is often called the "guardian of metabolism." Turns out you'll be doing just that with the combination of our strategies.

> ### MAXIMIZING ANTIAGING, PRO-LONGEVITY ACTIONS IN THE BODY
>
> ➤ Time-restricted feeding
> ➤ Exercise
> ➤ Resveratrol
> ➤ Omega-3s
> ➤ Curcumin
> ➤ DIM
> ➤ Pomegranate (ellagic acid)
> ➤ Milk thistle
> ➤ Carnosine
> ➤ Fisetin (in strawberries)
> ➤ Reishi mushrooms

Resveratrol

Resveratrol is one such secret tool in the kit. Jake Perry, a retired plumber in Austin, Texas, went down in feline history for breaking the Guinness World Record for raising the oldest cat—twice. The first record, from 1998, was for a part Sphynx, part Devon Rex named Granpa Rex Allen, who lived to age thirty-four; the second, from 2005, was a mixed tabby named Creme Puff who lived to age thirty-eight (that's more than twice the average life span of a cat!). His secret? In addition to a diet of commercial cat food with home-cooked toss-ins like eggs, turkey bacon, and broccoli, he did something else decidedly unusual: Every two days, he administered an eyedropper full of red wine to "circulate the arteries." Could the little dose of resveratrol in the wine have had big effects on the cats' longevity? Jake thinks so, and while we won't endorse alcohol *whatsoever* for pets, there's something to be said about this well-documented ingredient. (You've probably heard of resveratrol; this is the same polyphenol naturally present in grapes, berries, peanuts, and some vegetables; it's what gives red wine its health halo.) Obviously, we don't feed dogs grapes, but there is a safe source of resveratrol awaiting our canine companions.

For pet supplements, resveratrol is derived from the root of the Japanese knotweed (*Polygonum cuspidatum*), which is a rich source of the antioxidant that is used extensively in traditional Japanese and Chinese medicine.

It's just begun to make a splash in the dog world where it's been shown to have anti-inflammatory and antioxidant effects and also to confer anticancer and cardiovascular benefits, boost neurological function, aid in improving alertness in dogs, and reduce risk for all manner of mind-related ailments, from depression to cognitive decline and dementia.

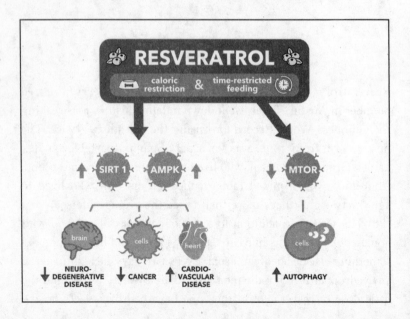

Dose: The dose for Japanese knotweed ranges from 5 to 300 mg/kg daily in dogs, with the high range being studied for its benefits with hemangiosarcoma. Over-the-counter dog products contain very low concentrations. A middle-of-the-road well-being dose studied in animals is 100 mg/kg per day, divided on food.

Curcumin

Looking for the Swiss Army knife of supplements? As noted in the previous chapter, curcumin is both a therapeutic agent used for a spectrum of health conditions and nature's anti-inflammatory. The compound targets biochemical pathways associated with neurodegenerative disorders that include cognitive impairments, energy/fatigue, mood, and anxiety. It's also a potent antioxidant, hormonal and neurochemical modulator, helper in fat metabolism, warrior against cancer, and a general friend to the genome. And it's high in fiber and rich in vitamins and minerals. Grating fresh turmeric over your dog's food daily is a great idea, but most people find adding the super-concentrated supplement to be far more beneficial.

Dose: 50 to 250 mg two times a day (roughly 2 mg per pound of body weight twice a day).

Probiotics

Several probiotic formulas are on the market for dogs. Look for one that contains a blend of different probiotic species with a large CFU (colony forming units) that's been third-party validated for viability and potency. We recommend rotating through many different brands and types of probiotics: soil-based (or spore-forming) and bacterial strains all have different attributes that help diversify your dog's gut microbes. You'll also want to add some prebiotic foods found on page 235 to arrive at the golden finish line: "postbiotics." Postbiotics come from rich sources of polyphenols that must be supplied in the diet and are sensitive to heat, which is another reason why ultra-processed pet foods are suboptimal. Fermented veggies and plain kefir are excellent food sources of probiotics, but many dogs won't tolerate their tangy taste. Use food sources of probiotics if your dog will eat them; if not, rotate through a variety of different dog-formulated probiotics (different brands and types) to nourish the microbiome. Follow the specific instructions on the package for each product. Blends of probiotics and digestive enzymes also can be very beneficial for many dogs.

Prebiotic foods + probiotics (fermented foods or supplements) = postbiotics. Postbiotics are now recognized as beneficial for dog health and well-being.

Essential Fatty Acids (EFAs)

All fatty acids are essential to the structure and function of cell membranes, especially in the brain. Studies have looked at people with extremely high levels of omega-3 fat, a type of polyunsaturated fatty acid (PUFA), and compared them with people who exhibit far lower levels. The results show that the higher levels are correlated with better memory and a larger brain. For dogs, the science is *very clear*: Fish oil improves skin, behavior, and brain and heart health; makes puppies smarter; and reduces inflammation and epilepsy. Without fatty acids, cells would simply fall apart. Cellular membranes are lipid envelopes that encase and protect the internal workings of cells. The membrane is essential for the production of energy in the mitochondria, because without the double membrane structure there is no storage space for the separation of electrical charge—no way of conducting chemical reactions to create energy.

The volume of cellular membrane in the body is mind-boggling. And your dog's requirements for EFAs are profound. The challenge is they must be obtained through the diet, because dogs can't make them if their diets are deficient. If the food you are feeding your dog has been heat processed, the amount of EFAs in the food has been affected, which is why we recommend supplementing.

Add supplements that provide more superstar omega-3s: eicosapentaenoic acid (EPA) and docosahexaenoic acid (DHA). These fatty acids, which are the preferred form of omega-3 fatty acids for dogs and often are derived from fish or a marine, ocean-sourced oil (salmon, krill, squid, mussels, etc.), have been shown to reduce inflammation and boost brain regeneration (including increasing canine BDNF). The real superheroes are the substances contained in the marine oils, the resolvins. These compounds block inflammation from occurring and resolve existing inflammation. Other types of healthy oils and fats (including hemp, chia, and flax oil) do not contain resolvins or DHA and EPA. The problem is these delicate compounds are inactivated with heat.

Due to high-heat rendering and pet food processing, most of the essential fatty acids in processed pet food have been destroyed. Just when you thought your bag of dog food has a good ratio of omega-3 to omega-6, get this: Any remaining omegas can easily break down once you open the bag. Another reason why it helps to add a stable, high-quality supplement to this madness, and why we are always suggesting that you add extra omega-3s on top of your normal pet's diet. (Note: Keep your dog food in your freezer to slow the degradation.) But get your EPA and DHA from the ocean and not plant sources, which don't contain adequate amounts for dogs. Ocean-sourced omegas are the most bioavailable and can be sustainably sourced and third-party tested for contaminants.

The confusion (and negative press) around fish oil supplementation lies in the form of the fish oil. Fish oil supplementation has come under scrutiny from many studies demonstrating that the more refined form, ethyl ester (which is cheaper to produce than the naturally occurring triglyceride or phospholipid form), can rapidly oxidize and deplete the body of antioxidants (not the goal). When you buy fish oil, make sure it's the triglyceride or phospholipid form. We rotate through a variety of oils sourced from salmon, krill, anchovy, mussels, and squid. If your dog is allergic to ocean-sourced oils, which is rare, vegetarian sources of high-DHA microalgae oil (algal oil) can be an alternative (not microalgae powder, which doesn't come close to meeting DHA or EPA requirements).

If your dog is eating fresher food that hasn't been high-heat processed and hasn't been sitting on a shelf for a year, you can supplement with fewer omega-3s. If you use fatty fish, such as sardines or cooked salmon as Core Longevity Toppers (CLTs) three times a week, you don't have to supplement at all!

Dose: According to board-certified veterinary nutritionist Dr. Donna Raditic, the anti-inflammatory effects of EPA and DHA in dogs with various diseases—including kidney disease, cardiovascular disorders, osteoarthritis, atopy (skin conditions), and inflammatory bowel disease—have been evaluated, and the dose ranges from 50 to 220 mg/kg (23–100 mg/lb) of body weight. The highest dose is recommended for osteoarthritis in dogs not otherwise being supplemented (not getting additional omegas in commercial dog food). If your dog isn't eating any other sources of omega-3s (like sardines), consider a maintenance dose of 75 mg per kg (34 mg/lb) of body weight for healthy animals. These doses are based on adding the milligrams of EPA and DHA per supplement capsule or milliliters of liquid. We recommend storing omega-3s in the refrigerator once opened; aim to use up your supply within thirty days, or buy capsules and hide them in a meatball. (Or puncture capsule and squeeze into food.)

Note: Cod liver oil is liver oil (not body oil) and can be high in vitamins A and D but not high in omega-3s. Some recipes call for cod liver oil as a source of fat-soluble vitamins. We don't recommend adding cod liver oil to your dog's diet unless it's an ingredient in a food recipe or your dog's blood tests show he is low in vitamins A and D.

Many Americans and people living in northern hemispheres have low levels of vitamin D, and some breeds do, too, despite eating fortified diets. Supplementing dogs with extra fat-soluble vitamins (particularly vitamins A and D) can create toxicities quickly, so never give vitamin D without asking your vet to first measure your dog's levels. Research shows northern breeds ("snow dogs") need more of vitamins E and D, omega-3s, and zinc to avoid nutritional dermatosis (skin problems). It's easy to hear about this and begin supplementing, but it's a recipe for disaster. If you think your dog is lacking in a specific mineral, ask your vet to check before you supplement.

Quercetin

This gem (pronounced KWAIR-suh-ten) smashed the Internet when we first wrote about it on Planet Paws. All it took was a callout for pet parents with allergy-ridden dogs—yeasty ears, runny/goopy/red eyes, itchy/scaly skin, and sneezing (among other symptoms of environmental allergens on the prowl). Vets view quercetin as nature's Benadryl, because it's well known to help with dog allergies. Quercetin is an important dietary polyphenol that is present in several foods and is consumed almost daily. It is a naturally occurring polyphenolic flavonoid that is commonly found in different fruits and vegetables such as apples, berries, and green leafy vegetables. Research in the bioactivity of quercetin as a strong antioxidative, anti-inflammatory, anti-pathogenic, and immune regulator has identified numerous pathways by which this phytochemical powerhouse may prevent or slow down the development of degenerative diseases in addition to its natural antihistamine qualities.

Apart from its antioxidant and anti-inflammatory properties, quercetin has been shown to help control mitochondrial processes that are likely to affect the entire cell and tissue. And new science shows that quercetin supplementation may have beneficial effects on neurodegenerative diseases in particular: In mice models that mimic Alzheimer's disease, it reduces the adverse buildup of the plaque proteins associated with the disease. It also inhibits AGE formation in the body. Bonus: The molecule may also reduce zombie cells.

Dose: Multiply the weight of your pet in pounds by eight (e.g., a fifty-pound dog should get 400 mg per day, a 125-pound dog would get 1,000 mg per day—the equivalent of consuming 124 red apples or 217 cups of blueberries). Tip: Whatever amount you give your pet, always split the dosage into two separate portions throughout the day; for best results, hide capsules or powder in their meals or a treat. If your pet is in really bad shape, you can always double the dose for this particular supplement.

TIP: ALMOND BUTTER IS THE FRESHER "PILL POCKET"

A dab of organic raw almond butter (thirty-three calories per teaspoon) is an easy substitute for ultra-processed semimoist treats made to hide pills. We make our own by whizzing up fresh, organic almonds in a food processor. In addition to reducing oxidative stress, almonds can significantly reduce the level of C-reactive protein in people, and they contain lignans and flavonoids. Raw, organic sunflower seeds can also be ground to make a DIY vitamin E-rich spread that hides pills and powders for finicky dogs. You can also use a tiny meatball, fresh cheese (proven to build a dog's microbiome), or a dollop of 100 percent pure pumpkin (freeze the rest in ice cube trays for future use). Peanut butter can be contaminated with mycotoxins, and some brands contain xylitol, which is toxic to dogs.

Nicotinamide Riboside (NR)

Ask anyone in the antiaging biotech world about the most promising molecules to extend longevity and this beauty surely will be mentioned. Nicotinamide riboside (NR) is an alternative form of vitamin B3 and a precursor to nicotinamide adenine dinucleotide, or NAD+, a star molecule that acts as a coenzyme in many critical processes in the mammalian body, including cellular energy production, DNA repair, and sirtuin activity (enzymes involved in aging). Without NAD+ acting as a coenzyme, these processes simply cannot occur and life would not exist. It's so important that it's found in every cell of the body. But mounting evidence suggests that NAD+ levels decline with age—a change that scientists now consider a hallmark of aging. Lower NAD+ levels also are responsible for many age-related conditions like cardiovascular disease, neurodegenerative diseases, and cancer.

For example, a study in aging mice demonstrated that supplementing orally with nicotinamide mononucleotide (NMN), another larger NAD+ precursor molecule, prevented age-associated genetic changes and improved energy metabolism, physical activity, and insulin sensitivity. It's not easy to boost levels of NAD+ because in supplement form, NAD+ has very poor bioavailability, but NR is a great way to boost natural levels. Animal studies have shown that supplementing with NAD+ precursors NMN or NR restores NAD+ levels and slows age-related physical decline. Most antiaging experts we consulted admitted that they take NR or NMN daily. Interestingly, when NMN was given to Beagles experimentally, and we started taking it, too, it also reduced lipid and insulin levels.

Dose: There's a huge dose range, with many products for humans suggesting 300 mg a day (roughly 2 mg per pound of body weight for your dog). Animal studies indicate much higher doses (32 mg per kg of body weight a day) offer greater benefits, but because this supplement is so expensive, give when you can afford, starting with 2 mg per pound of dog.

Nootropics: Sometimes called "smart supplements," nootropics are compounds that enhance brain function by helping to prevent or slow cognitive decline. Studies have revealed that people experiencing cognitive impairment are deficient in essential vitamins and nutrients that play a protective role against cognitive decline, the same holds true for animal models. Scientists discovered that specific nutrients play an important role in cellular activity, which is necessary to maintain optimal cognitive function. Research has shown that chronic stress can accelerate cognitive decline and impair memory function. Some nootropics contain ingredients that are considered adaptogens, meaning they help your body cope with stress and, in turn, improve cognitive function.

Lion's Mane Mushrooms

This nootropic mushroom has a wide range of cognition-enhancing effects, including being a potent adaptogen (substances that assist the body in healthfully adapting to stress). Research also shows promising improvements for depressive and anxiety behaviors in animal models. A beneficial polysaccharide in the mushroom proves useful for treating and preventing GI issues, including ulcers, and reduces nervous system injuries and degeneration in animals. We especially love it for protecting the myelin in breeds at higher risk of degenerative myelopathy. It's a rich GI protectant, and it improves the immune system in the intestines, allowing the gut to ward off ingested pathogens. Feeding fresh lion's mane mushrooms as CLTs is fantastic, if you can find them. If you can't, or if your dog won't eat them, consider supplementing, especially in pets over the age of seven.

Dose: In one Japanese study on cognition, humans received a total of 3,000 mg a day with beneficial results; this translates into 1,000 mg for every fifty pounds of your dog's body weight.

Glutathione

We've already covered this vital amino acid that is produced by the body and is pivotal in the breakdown of carcinogens. Glutathione helps remove harmful AGEs from ultra-processed foods, neutralizes free radicals, and detoxifies industrial and veterinary toxins. It also can help protect a dog from heavy metal damage. In your dog's liver, detox pathway activities that entail glutathione account for up to 60 percent of the toxins created in the bile (bile is your dog's liver's main vehicle to get rid of excessive substances). That's why glutathione is referred to as the master antioxidant. Glutathione also recharges other antioxidants, strengthening their ability to fight inflammation; it serves as a cofactor for dozens of enzymes that neutralize damaging free radicals. In studies, clinically ill dogs have less glutathione. Using a blend of medicinal mushrooms as CLTs is ideal, but if your dog won't eat mushrooms, supplementing with glutathione is a good idea, especially as your dog ages.

Dose: There's a wide swing in glutathione dosing, but most doctors recommend 250–500 mg a day for healthy humans, or 2–4 mg per pound of body weight for your dog, popped in a meatball or a treat in between meals.

Are dog dementia drugs effective? Yes, they work. Low-dose deprenyl (selegiline) is the only FDA-approved treatment for canine cognitive dysfunction and is most known for stimulating production of dopamine, an important neurotransmitter involved in emotions, pleasure sensations, and the brain's reward and motivation mechanisms. Dopamine also helps control movement. Deprenyl is used in veterinary medicine to block the enzymatic activity of a substance that slows down the breakdown of neurochemical dopamine. Selegiline increases neurotrophic factors, compounds that strengthen existing neurons, and supports the growth of new neurons. It also increases a powerful antioxidant that breaks down harmful substances themselves. This helps prevent tissue damage that can lead to hardening of the arteries, heart attack, stroke, coma, and other inflammatory conditions. If you want to try this drug, talk to your vet about it; we recommend beginning it in combination with a complete lifestyle change, as early as possible, when canine cognitive disorder has been diagnosed. Doctors have known about the longevity benefits of selegiline since the 1980s. Even back then a handful of animal studies showed that selegiline leads to measurable life-span increases.

Custom Curated Support

If you use a lot of chemicals in your home or yard, add SAMe to your dog's diet. If you use heartworm, flea, or tick medications, add milk thistle.

SAMe: S-adenosyl methionine (SAMe) is a naturally occurring molecule made in the liver of dogs and functions as a methyl donor for a variety of compounds required for detoxification. SAMe is necessary to repair your dog's DNA via methylation and is also a precursor to many key biomolecules including creatine, phosphatidylcholine, coenzyme Q10 (CoQ10), and carnitine. All of these chemicals in the body play a role in pain, depression, liver disease, and other conditions. SAMe also participates in the production of many proteins and neurotransmitters, and has been approved as a nutraceutical since the 1990s (because it isn't found in food, supplementing with SAMe is sometimes advisable). Numerous double-blind studies have proven its efficacy in relieving depression and anxiety. And human clinical trials have shown it to be as effective as nonsteroidal anti-inflammatory drugs (NSAIDs), making it a good candidate for decreasing pain and reducing swelling. In the canine world, vets use SAMe to help treat cancer, liver problems, and canine cognitive dysfunction syndrome.

One popular brand of SAMe for dogs reported a 44 percent reduction in problem behaviors, including a reduction in house soiling, after both four and eight weeks (compared to 24 percent in the placebo group). Other documented benefits included marked improvement in activity and playfulness; significant increase in awareness; decreased sleep problems; and decreased disorientation and confusion. Another unrelated study on laboratory dogs showed improvement in cognitive processes, including attention and problem solving. There are many veterinary-prescribed brands of SAMe, but you can also buy it over the counter. Give 15–20 mg per kilogram of body weight, once daily. This supplement is best absorbed if not given with a large meal, so pop it in a meatball and give it in between meals.

Milk thistle (silymarin) is the go-to liver detox herb. Want to flush out lawn chemicals, air pollution, and residues from veterinary drugs, including flea and tick meds, heartworm medications, and steroids? One of the most important herbs to add to your pet's diet, this rockstar toxin flusher comes from a flowering herb especially famous for tackling liver problems by serving as a housecleaner of sorts. Detoxification is such an important process, not only for humans but also for our pets. A dog that can't detoxify appropriately is at risk for serious immune complications downstream. Milk thistle is the boss of detoxifiers. According to University of Maryland Medical Center, "Early laboratory studies show that silymarin and other active substances in milk thistle may have anticancer effects. These substances appear to stop cancer cells from dividing and reproducing, shorten their life span, and reduce blood supply to tumors."

Dose: Give one-eighth of a teaspoon of the loose herb per ten pounds of body weight. This herb should be pulsed (given intermittently) for maximum efficacy. Give it daily for a week after heartworm pills (or to help clear other drug residues from the body) or for the week after your apartment complex applies lawn chemicals. Milk thistle is widely available in many pet-specific products. If you buy a human product, a common "detox" dose is 50–100 mg/kg (22–45 mg/lb) daily. Look for a minimum of 70 percent silymarin on the label.

For extra joint support for all ages—from young dogs with joint trauma to older dogs with degenerative joint disease—the perna mussel is a Longevity Food that is available in supplement form for the musculoskeletal system.

Perna mussels (aka "green-lipped mussels" [GLM]) work in a similar way to nonsteroidal anti-inflammatory drugs (NSAIDs). Hailing from New Zealand, the supplements are derived from mussels that have a bright green stripe around the shell and an additional green lip inside the shell. The Maori have a long history of using GLMs, and those living on the coast show much lower levels of arthritis than the Maori people living inland. Additionally, it has been proven that symptoms of osteoarthritis in dogs can be alleviated by using the extract of these mussels. A 2006 double-blind placebo-controlled study of eighty-one dogs who had mild to moderate degenerative joint disease found that dogs greatly benefited from long-term use of a 125 mg supplement of GLM. According to a 2013 study printed in the *Canadian Journal of Veterinary Research*, adding GLM to the diet of a dog with OA greatly improves their way of walking compared to those not given the supplement. These dogs also absorbed high levels of EPA and DHA omega-3 fatty acids into their blood. In conclusion, dogs with OA were greatly helped by GLM extract. Easy ways to incorporate GLM into your dog's diet is through freeze-dried mussels as dog treats or using powdered supplements.

Dose: 15 mg/lb (33 mg/kg) of body weight daily, divided in meals.

For dogs that need extra support for stress and anxiety (always in addition to behavior modification and daily movement therapy):

L-theanine is a calming amino acid found primarily in tea. It promotes alpha wave brain production, which reduces anxiety and noise phobias, and supports a relaxed but focused mindset. Veterinary-prescribed products are available, but L-theanine is also widely available in human health food stores. The most effective dose in reducing anxiety in dogs is 2.2 mg/kg (1 mg/lb) of body weight, given twice daily.

Ashwagandha, a small evergreen shrub that grows in India, the Middle East, and parts of Africa, is called an "adaptogen" because it helps the body deal with stress by supporting brain function, lowering blood sugar and cortisol levels, and helping fight symptoms of anxiety and depression. It's also been shown to help improve liver function in older dogs. Dose: 50–100 mg/kg (23–45 mg/lb), divided into two doses on food.

Bacopa monnieri is a staple plant used in Ayurvedic medicine, and numerous clinical studies have found that it enhances memory retention—dogs learn faster and remember longer—and reduces stress and anxiety, including depression. Some animal studies have even shown that its anti-anxiety ("anxiolytic") efficacy is comparable to benzodiazepine (e.g., Xanax) but won't make your dog sleepy. Doctors around the world are adding bacopa monnieri to their support protocols for patients with cognitive degeneration because it's a proven memory enhancer.

Dose: 25–100 mg/kg (11–45 mg/lb) a day on food, divided in meals. Use the low end for cognitive well-being and the high end for addressing anxiety.

Rhodiola rosea is another adaptogenic herb that fortifies the body to better handle stress. Several studies have found that rhodiola rosea supplements improve mood and decrease feelings of anxiety.

Dose: 2–4 mg/kg/day (1–2 mg/lb) divided in meals should do the trick.

For dogs spayed or neutered early in life (before puberty), lignans can be helpful in balancing remaining hormones if a dog has been desexed in the first year of life. Lignans are phytoestrogenic, meaning they are plant-based compounds that mimic estrogen in the body, which send feedback to the adrenal glands to stop producing inappropriate amounts of estrogen (which was never their job in the first place!). There are myriad sources of lignans, including flax hulls (not to be confused with flaxseed, which don't contain enough lignans), cruciferous veggies, and the knots of spruce pine trees (HRM lignans). Lignans are routinely prescribed in veterinary medicine as an adjunct support for dogs with Cushing's disease (excessive production of adrenal hormones). Significantly elevated ALP (alkaline phosphatase) is a common clue on routine blood work that cortisol may be high and should be checked. Lignans are often used in combination with melatonin and diindolylmethane (DIM) in several "dog hormone balance" products to reduce cortisol and help alleviate stress in adrenal glands, which work overtime after desexing surgery. Use 1–2 mg of lignans per pound (2.2–4.4mg/kg) of body weight a day.

For dogs fed more than 50 percent ultra-processed food: According to research, you can safely assume your dog has elevated levels of AGEs. If she's not eating organic food, she could also be harboring detectable levels of pesticide residues, and possibly heavy metals and other contaminants (such as PBDEs, phthalates), so you need to provide a means of helping his body clear them out. Some gems we like:

Carnosine is a protein building block that is naturally produced in the body in small amounts and has been demonstrated to help prevent the body from absorbing and metabolizing AGEs and ALEs (advanced lipoxidation end products from fats—another by-product of ultra-processed foods you don't want accumulating). Carnosine demonstrates natural antioxidant protection, heavy metal chelation, and the ability to detoxify reactive molecules generated from ALEs and AGEs and *it inhibits formation of them*.

Dose: We recommend 125 mg a day for dogs less than twenty-five pounds, 250 mg a day for dogs up to fifty pounds, and 500 mg a day for dogs that weigh more than fifty pounds. This human supplement is available at your local health food store or online.

Chlorella, a single-celled, freshwater medicinal algae, binds heavy metals and food and environmental contaminants. You can supercharge chlorella's superpowers by feeding it with cilantro, which removes glyphosate residues found in ultra-processed dog food and conventionally grown produce. Chlorella is a human supplement that works well for dogs because it comes in tiny tablets you can hide in a meatball or powder you can mix in food.

Dose: 250 mg a day for dogs under twenty-five pounds, 500 mg a day for dogs up to fifty pounds, and 750–1,000 mg a day for larger dogs.

CHEMICAL DETOX SUPPLEMENTS

➤ Veterinary and environmental pesticide cleanup:
  ➤ milk thistle, SAMe, glutathione

➤ Mycotoxin, glyphosate, and heavy metal cleanup:
  ➤ quercetin, chlorella

For Dogs with Chronic Infections

Olive leaf extract: The extract from leaves of the olive plant—not from the fruit itself—is equally and even perhaps more potent than the oil, as it contains an active ingredient called "oleuropein" that is thought to contribute to the anti-inflammatory and anti-oxidant properties. Oleuropein has beneficial effects on maintaining healthy blood sugar levels in dogs, contains polyphenols that target brain cell longevity in animals, and induces autophagy via the AMPK/mTOR signaling pathway, in addition to protecting against a number of common pathogens and parasites. Oleuropein also has strong antimicrobial and antiparasite properties, prevents and treats liver disease and toxicity in a number of animal models, and is being studied for its effectiveness against neuro-degenerative diseases; it also kills senescence cells and stimulates Nrf2. This potent polyphenol induces powerful apoptosis and is being trialed on many aggressive cancers due to its inhibition of abnormal cell growth.

Dose: Look for a minimum of 12 percent oleuropein in the human herbal product you buy: 125 mg twice daily for dogs less than twenty-five pounds, 250 mg twice daily for dogs up to fifty pounds, and 500–750 mg twice daily for larger dogs. Use for six to twelve weeks to assist in controlling active infections (particularly recurrent skin, bladder, and ear infections) and to stimulate autophagy, then cycle off for three to four weeks before reinstituting.

Our Favorite Senior Dog Supplement

Ubiquinol is the active form of coenzyme Q10 (CoQ10), a fat-soluble, vitamin-like antioxidant the body needs to support and maintain natural energy production inside the mitochondria of cells, helping them to function at optimal levels. No surprise: The heart and liver contain more mitochondria per cell than other body parts and thus contain the most CoQ10. Also no surprise: CoQ10 is one of America's most popular supplements for human consumption and is recommended for cardiac patients, both as a treatment and for prevention of age-related cardiac diseases. In the vet world, canine cardiac patients are prescribed this supplement to slow the progression of congestive heart failure. In one of the first studies to evaluate dogs suffering with mitral valve disease (MVD), the most common heart disease found in small breed dogs, CoQ10 significantly improved cardiac function in small dogs. We also recommend it prophylactically, to nourish aging mitochondria and reduce the likelihood of cardiovascular disease. It's impossible to get enough CoQ10 through diet alone. Ubiquinol (the more bioavailable form of CoQ10) is a much more expensive supplement but is more easily assimilated.

Dose: Ubiquinol doses vary from 1–10 mg/pound once or twice a day, depending on your health goal. A once-a-day dose is sufficient to maintain mitochondrial well-being and heart health. For animals with cardiovascular disease, dose twice a day. Note: Oil-based preparations of ubiquinol are considered more effective and more easily assimilated than powdered sources of regular CoQ10. Oil-based ubiquinol is sold in soft gel caps or a liquid pump, while crystalline CoQ10 is sold in capsules, tablets, or powders. Tip: If you buy plain CoQ10, use the higher dose suggestion for maintenance health and give it with a teaspoon of coconut oil for optimal absorption.

I (Dr. Becker) met Ada as a puppy in 2004. My first wellness goal was to intentionally create a gut of steel because a healthy gut translates into a healthy immune system. Genetically, her Pit Bull DNA would lean toward expressing atopic dermatitis (allergy-like symptoms similiar to eczema), which I wanted to avoid. I spent my days at my animal hospital working with owners that were at the end of their ropes, desperate to avoid euthanasia. Like many functional medicine doctors, I was the last stop for animals with incurable maladies: allergies, cancers, musculoskeletal issues, and organ failure. The last thing I wanted was to come home at night to a miserably itchy dog, but I knew this would take an intentional epigenetic alteration plan, as I like to call it (and a worthy topic for a separate book).

As we explained in the first two parts, our dogs carry DNA that may or may not be expressed, depending on epigenetic factors that are influenced by our dog's environment. I was well aware, as her guardian, that I held great power in keeping her itchy genes at bay or allowing "nature to takes its course" through her heritable atopic (itchy) predisposition. My intention was to purposefully decrease her likelihood of expressing atopy-prone DNA. I started by creating and protecting a healthy microbiome. I did not deworm her "just because"; instead, I checked a stool sample every month for three months to confirm she was parasite-free. She came to me eating 100 percent ultra-processed puppy food. I immediately started her on different brands of dog probiotics with different strains of beneficial bacteria, adding a pinch of one or another at each meal. I was too busy to make all homemade meals for her, but I immediately weaned her to a variety of nutritionally complete raw food brands, rotating through a different protein source (and brand) for each meal. I had two giant freezers, so storing a variety of small bags of dog food wasn't hard. Rotating through meals of beef, chicken, turkey, quail, duck, venison, bison, rabbit, goat, emu, ostrich, elk, salmon, and lamb (along with all the different veggies in each formula) fostered nutritional and microbial diversity early on.

Ada had daily exposure to healthy soil (I lived in the forest) and spent lots of time outside. I live a pretty "green" lifestyle, so her exposure to household and environmental chemicals was minimal. I was hell-bent on not giving her antibiotics unless required for some life-threatening reason (I knew then that it takes months for a dog's gut to restore itself, even after a short course of antibiotics). She had the inevitable "puppy pyoderma," the belly and body acne many puppies get as their maternal antibodies wane and their own immune system kicks in. This is a common time for puppies to receive their first round of unnecessary antibiotics. I managed her breakouts by dabbing her pimples and pustules with povidone iodine twice a day. During that time, I also used olive leaf as a supplement, for one month. She had several bouts of diarrhea from eating stuff she shouldn't have, as most puppies do. I managed her diarrhea without GI antibiotics. (Flagyl, or metronidazole, the most prescribed antibiotic for GI problems, effectively treats the symptoms of diarrhea but equally effectively creates dysbiosis, the first step in the atopy equation.) By giving her activated charcoal three times a day on an empty stomach, as well as a few meals of cooked fat-free turkey mixed with canned pumpkin (plus slippery elm), her "dietary indiscretion" always resolved in a timely fashion.

Ada came into my life already having received two puppy vaccines. Instead of automatically giving her more vaccines, I wanted to determine if she was sufficiently immunized to protect against life-threatening viruses, long-term. A simple blood test, called an "antibody titer test," revealed she was indeed already protected; giving her more puppy "booster shots" wouldn't benefit or "boost" anything. Over the years, her titer tests continued to demonstrate protective immunity from her initial puppy vaccines, even sixteen years later.

I tailored her supplements to her body's specific needs during different life stages. When she was a young dog, I wanted to protect her tendons and ligaments (another weakness in the breed). When she was a midlife dog, I wanted her immune system to be resilient. When she became a senior, I wanted to protect and preserve her organ function; and now that she's a geriatric dog, my focus is to slow cognitive decline and manage any discomfort in her body. At seventeen, she also needs eye support. For me, medicine is as much an art as a science. The art is to customize a patient's dynamically changing wellness protocol over time, factoring in genetics, and as dictated by the patient's specific health needs rather than prescribing a standard "one size fits all for life" protocol. As your dog's body changes, so will your supplement regimen.

WHAT IS A FUNCTIONAL MEDICINE DOCTOR?

Functional medicine considers food and lifestyle to be the primary modes of healing, rather than pharmaceutical interventions as the first or only option for managing chronic disease. Functional medicine veterinarians strive to identify and remove lifestyle and environmental obstacles before disease occurs. We create customized, dynamic wellness protocols for animals, with the goal of promoting a higher state of ongoing well-being, an above-average quality of life, and an above-average life span. This differs from the conventional medical approach of reactively treating ailments and diseases after symptoms alert us the body is diseased or degenerating. Find a list of professional animal organizations that embrace functional medicine on page 408.

We could write a separate encyclopedia of dog supplements, there are so many brands and beneficial nutraceuticals and herbs clinically proven to improve health. Other vets have attempted this, but what's most important (in terms of not flooding the bowl with excessive pills, making sure you know what you're giving and why, and not spending a fortune) is the wise assessment of which supplements best serve the individual. Like humans, different animals need different support at different times and for different reasons. Partnering with a functional or wellness veterinarian or consulting a vet that focuses on proactively preventing disease from occurring can be really helpful. **We also encourage pet guardians to become knowledgeable in order to best advocate for your animals**.

This industry changes fast, too. In recent years, for example, CBD (cannabidiol) products for dogs have flooded the market. CBD is a compound found in cannabis and hemp. Most CBD products, especially those designed for dogs in the form of oils and tinctures, are derived from hemp and not from marijuana, the latter of which also contains tetrahydrocannabinol (THC), the compound that gives marijuana its psychoactive properties. As a wellness supplement, CBD is touted as a panacea of sorts with multiple beneficial effects on the body; it acts as an anti-inflammatory, calms the nervous system, treats pain and anxiety, and even potentially prevents and adjunctively treats cancer. And while we've experienced the benefits of using this herb to manage specific challenges with our own dogs, the biggest issue we see with dog CBD products on the market (besides quality control and potency issues) is the misguided assumption that it can manage all sources of physical pain and improve every behavioral issue. It won't. CBD and many other herbal products may be beneficial for your dog in certain circumstances, therapeutically. The supplements we've listed here fall into the "wellness" category—supplements you can use on a daily basis, if you wish, to incrementally improve health and delay aging.

If your dog has a specific health challenge, many nutraceutical pro-tocols can be incredibly beneficial when customized around your dog's unique medical issues and physiology. Wellness companies are beginning to offer supplement protocols customized around your dog's specific predispositions, DNA test results, and unique issues.

If your dog is unwell or on medications, talk to your vet about the supplements you'd like to start using. Always tell your vet what supplements your dog is taking prior to surgery or beginning new prescription medications. Supplements can be mixed in with food or hidden in a small meatball or a dab of almond butter or fresh cheese (did you know research shows fresh cheese containing probiotics can benefit our dogs' microbiome?!). Never force loose powders down your dog's throat; it breaks trust, creates a choking hazard, and doesn't feel good.

LONGEVITY JUNKIE TAKEAWAYS

➤ The right combination of supplements at the right time—without going to extremes—can help support your dog's natural biology to address any gaps in their diet and other lifestyle factors, age, or genetics. But not all dogs need supplements all the time.

➤ Core basics (see chapter details for dosing and delivery) to consider for your dog:
  ➤ Resveratrol (knotweed)
  ➤ Curcumin (especially if your dog won't eat turmeric)
  ➤ Probiotics (especially if your dog won't eat fermented veggies)
  ➤ Essential fatty acids (EPA + DHA, if your dog doesn't consume fatty fish two to three times a week)
  ➤ Quercetin

- ➤ Nicotinamide riboside (NR) or nicotinamide mononucleotide (NMN)
- ➤ Lion's mane mushroom (for dogs over seven)
- ➤ Glutathione (if your dog won't eat mushrooms)

➤ Curated Support:

- ➤ For dogs exposed to a lot of chemicals (e.g., lawn-care products, household cleaning agents), add SAMe (also see box on page 261).
- ➤ For dogs given heartworm, flea, and tick medications, add milk thistle.
- ➤ For dogs who need extra joint support, add perna mussel.
- ➤ For dogs who need extra support for stress and anxiety, add L-theanine, ashwagandha, bacopa monnieri, and rhodiola rosea.
- ➤ For dogs spayed or neutered before puberty, add lignans.
- ➤ For dogs fed more than 50 percent processed food, add carnosine and chlorella.
- ➤ For dogs with chronic infections, add olive leaf extract when infections flare.
- ➤ For senior dogs, add ubiquinol.

9

Personalized Meals as Medicine

Pet Food Homework and Fresher Percentages for a Durable Dog

> The food you eat can be either the safest and most powerful form of medicine or the slowest form of poison.
>
> —Ann Wigmore

How much you can improve your dog's well-being depends on three factors: your belief that lifestyle matters (basically your commitment to the process, i.e., effort), genetics, and budget. While we can't alter the DNA that makes up your dog's genetic constitution, oftentimes we can influence their enzymatic pathways, epigenetically, through environmental changes, including diet. Our viewpoint about epigenetics is that all dogs have to eat: They may as well be eating foods that contribute positively to their genome's healthy expression.

It's important to check with your vet before instituting any lifestyle or dietary changes to ensure there are no underlying issues that need to be supervised during the process of transitioning to a healthier lifestyle.

Prelude to Powerful Change

Life's subtle influences become unknowing habits. Begin thinking about reshaping old patterns of behavior that aren't healthy and starting new, healthier habits. Start by evaluating your dog's current diet and deciding whether you want to make any significant changes. We recommend shifting food and treats slowly, to avoid GI upset; the planning part, conceptually, is important.

Remember, the goal of this program is to reduce metabolic stress and inflammation, activate AMPK and longevity pathways, assist your dog's body in clearing out accumulated toxins that may be stored in organs and tissues, and rebalance the microbiome.

We are going to assume you're probably feeding some processed or ultra-processed food right now; even if you aren't, keep reading to ensure you've assessed your recipes or brands against Forever Dog criteria. The criteria provide a way to assess the food

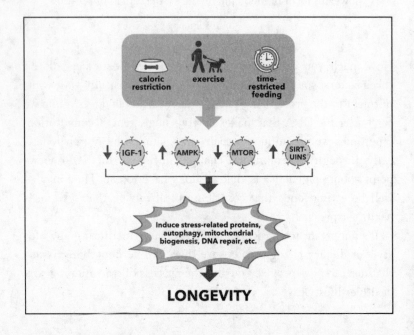

you are currently feeding (to decide if you want to continue or improve your base diet) and a template for choosing fresher brands, which can be included in any amount beginning now or anytime in the future.

Getting Started with Food Changes

For the sake of simplicity we've broken our approach to introducing healthier foods into two steps. The first step is to introduce Longevity Foods, using them as treats and as Core Longevity Toppers (CLTs). The second step is to improve your dog's daily diet, if it's necessary and you are willing and able to do so. The reason for making food changes in two steps is simple: It's less overwhelming for you and your dog. Slow food transitions of any type are less stressful for your dog, and give you time to do your research, complete the Pet Food Homework, and begin the enjoyable process of discovering your dog's gustatory preferences.

Up until now, you may have assumed that, because your dog eats his food, he *likes* it. You're going to discover that your pup has fine-tuned food preferences, likes and dislikes, just like you. She's never had the opportunity to discover and enjoy a variety of nutritious and delicious foods. Through trial and error and lots of opportunities to try tiny bites of new foods, you'll begin the thoroughly enjoyable (and oftentimes amusing) journey of discovering the intricacies of your dog's amazing nose and taste buds. The world is full of literally lifesaving foods for you and your dog to discover together!

Step 1: Introducing Core Longevity Toppers (CLTs)

10 Percent Core Topper Rule: Fresh food additions can be added in the form of CLTs to any brand of food you are currently feeding your dog. The great thing about the 10 percent rule is that these

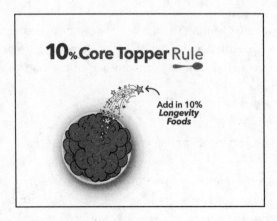

add-ins don't have to be nutritionally balanced—they're considered "extras," which, according to veterinary nutritionists, means 10 percent of your dog's total daily caloric intake can be freebies we can use to work longevity magic. As a reminder, if your dog is a little pudgy and could drop a few pounds, you can *replace* 10 percent of his food calories with CLTs; if your dog is lean and normal weight, you can *add* them in. You'll be working to add up to 10 percent of Longevity Foods in the form of CLTs, regardless of what food, or base diet, you feed your dog.

Step 2: Evaluate Your Dog's Base Diet and Freshen Up the Bowl

What is she eating for her daily nourishment? We want you to have complete clarity about your current dog food or any brand or type of dog food you may consider feeding in the future by doing three exercises:

1. **Pet Food Homework.** The results of these simple assignments will yield scores that provide the criteria you'll use to select the brands and diets to begin your Forever Dog Meal Plan, or to reinforce your confidence that you're feeding exactly what you intend to be feeding.

2. **Choose Your Fresh Food Category.** There are lots of choices, even within the category of less-processed dog foods. Here you'll review all the options and decide which best fits your dog's needs and your lifestyle (and you don't have to choose just one!).

3. **Set Your Fresh Percent.** Set a fresher-food consumption goal by choosing your dog's fresh percent—the amount of unprocessed or fresh, flash-processed dog food you'd like to work toward feeding at each meal. Put another way, the amount of ultra-processed pet food you'll be reducing or eliminating from your dog's life.

Improving the overall health of your dog's bowl begins with determining the base of your dog's daily sustenance, in light of your nutrition and health goals for your dog. Chances are you're very excited about getting to this point in the book because you finally get to put into action all you've learned in Parts I and II. You can start by asking yourself: How truly nutritious and health-promoting is my dog's food? What criteria am I using to come to my conclusion? You may not need to improve your dog's diet, but you'll still do your Pet Food Homework to affirm your dog is eating the best food you can afford to feed. Thousands of our clients and followers, after completing this exercise, recognized the disconnect between what they *thought* they were feeding and what they were *actually* feeding; there was room, sometimes lots of room, for substantial improvement.

Pet Food Math does a great job revealing areas you may want to address to maximize your dog's nutrient intake and minimize his consumption of unwanted tagalongs. In any case, we recommend doing what you can, at a slow and confident pace, and feeling genuinely good about what you *can* do. Every positive shift you make, no matter how small, results in a better health outcome, so don't compare yourself to others or allow guilt or frustration to creep in. And you certainly won't be doing everything at once, so relax and

enjoy mastering a skill that will serve you well: learning how to
evaluate dog food brands.

Exercise 1:
Do Your Good, Better, Best Pet Food Homework

————

Pet Food Homework allows you to evaluate the food you're currently feeding or any new brands of dog food you're considering
purchasing. If you don't intend to switch your dog's diet, we still
recommend you keep reading and apply these evaluation tools to
the food you're currently feeding; you can never know too much
about what's going into your dog's body on a daily basis. By the
end of this exercise, you'll apply objective criteria to rank pet food
brands as good, better, and best. Many people in our community
have completed this exercise and realized their beloved brand
doesn't make the cut—their brand fails miserably. Our response?
Now you know! (Thank goodness!) Now you have more information to make better choices. (Don't beat yourself up about stuff you
didn't know before.) Even if your brand falls to the bottom of the
"good" bandwidth (and far away from "better" or "best" levels), it
may nevertheless be your choice for your dog because it comports
with your personal food philosophy.

The purpose of this assignment is to gain a solid understanding of
a brand's *biological appropriateness*, the *amount of processing*, and *where
the nutrition comes from*. How important these topics are to you will
be based on your own opinions and values. If one area has a lower
score than another, that may be just fine with you, and that's okay.

Unfortunately, the data needed to develop an unbiased Consumer Reports–type website for dog food brands are not publicly
available; dog food companies rarely publish their internal research
or disclose their raw materials sourcing, and there isn't a National
Institutes of Health for Pets. "The List," an annual unbiased third-

party review assembled by Truth About Pet Food, is the best North America has. As you can imagine, however, it's a very short list compared to the hundreds of brands available on the market because it relies on companies willing to provide third-party documentation and sourcing transparency. This is where your own personal food philosophy comes into play. Along with your Pet Food Homework scores, you'll be armed with all the information you need to do your own brand assessments. Most people say, "Just tell me what brand to feed," or ask "Is X brand good?" It boils down to your definition of "good," right?

Ironically, both our fathers recited the same adage to us while growing up: "Give a man a fish, and you feed him for a day. Teach a man to fish, and you feed him for a lifetime." As endearing/annoying as it is to hear this (again), it certainly applies when choosing pet food brands. We teach you how to evaluate all types of pet food, so instead of asking, "Is this brand good?," you can say, "I feel confident in my choice of this brand of food for my dog." To get there, you need enough knowledge to make wise decisions, and we'll share that knowledge in the next section.

We don't recommend you automatically adopt another person's personal food philosophy as your own. Do some soul-searching and identify your own core food beliefs. What matters to you when buying food? What matters to you when buying dog food? Below are some considerations that have helped inform and shape the personal food philosophies of thousands of health-conscious pet guardians around the world. Use the questions on this list as a starting point for crystalizing your own core beliefs about each subject. Collectively, your opinions on these topics are what constitute your personal food philosophy about dog food:

Company transparency: Can I get honest answers about sourcing, quality of ingredients, and species-appropriateness?

Cost: Can I afford it?

Taste/palatability: Will my dog eat it?

Freezer space and preparation time: Can I store the amount of food I need, and do I have time to prepare the meal as it was intended?

Genetically modified organisms (GMOs): How important is it that my dog does not consume ingredients that have undergone intentional genetic manipulation?

Digestion/absorption testing: How important is it to know how well my dog can assimilate the food?

Organic: How important is it that my dog not consume Roundup or other pesticides and herbicides in the food?

Grass-fed/free-range: How important is it to avoid factory-farmed meats (and drug residues) or animals raised in concentrated animal feeding operations (CAFOs)?

Contaminant testing: How important is it that the food's raw ingredients are third-party tested for contaminants (i.e., euthanasia solution, heavy metals, glyphosate residues, etc.)?

Humanely raised/slaughtered: How important is it that the animals becoming "food" were not abused and did not die a grisly death?

Sustainability: How important is it that the food is manufactured in a way that maintains healthy ecosystems and minimally impacts the environment?

Nutritional testing: Does it matter whether the batch of food (or just the initial recipe) has been analyzed in a laboratory or feeding trial to demonstrate nutritional adequacy?

Synthetic-free: Does it matter whether my dog gets the bulk of nutrients from food versus lab-made vitamins and minerals?

Ingredient sourcing: Does it matter whether the food contains imported ingredients from countries that have different quality-control standards?

Quality of raw materials (feed or food grade): Does it matter whether my dog food is human grade (phrased another way: Does it matter that my dog food ingredients were rejected for human-food inspection)?

Nutrient levels: Does it matter whether my dog food meets minimum requirements for my dog to avoid nutritional deficiencies or excessive nutrients that could be harmful to my dog's health? Does it matter whether the company will share nutritional testing results with me?

Formulation: How important is it that my dog food meets a nutrition standard (NRC, AAFCO, FEDIAF) and does it matter who formulated the food?

Quality control: How important is food safety and product quality control?

Processing techniques: How important is avoiding the feeding of Maillard reaction products (MRPs) in dog food (including AGEs, ALEs, heterocyclic amines, and acrylamides)?

There are many other "food issues" that may shape your personal food philosophy, including many not listed here. Give careful consideration to each of these issues before you choose a food category and brand.

There's a company and type of food to fit almost every personal food philosophy, and there are always homemade recipes. (Visit www.foreverdog.com for some recipes to inspire you.) A lot of people told us they didn't know they *had* a personal food philosophy until they thought through these questions. Many were surprised and disappointed to learn the brands they loyally patronized for years were wildly incongruent with their personal food philosophies. Within each category of dog food there are bad, good, better, and best options. As your budget, life, and food philosophy evolve over time (and they usually do), you'll reassess and retool your Forever Dog Meal Plan. Companies sell, change hands, and reformulate their products; we suggest you do a review of the brands you're feeding on an annual basis. And we can't say this enough: **Hybrid meal plans, or rotating through a variety of dog food products from several different brands over the course of a year,**

are one of the best ways to protect against the pitfalls of monotonous diets.

If you opt to make homemade food, you're in complete control of the quality and source of ingredients you use, but if you are going to buy dog food, we make one solid recommendation that applies to all dogs, regardless of age, lifestyle, and geographic location: Avoid the Dirty Dozen list, if at all possible.

The Dirty Dozen: Avoid buying dog food with any of these ingredients listed on the labels (in no particular order):

➤ Any type of meal (i.e., "meat meal," "poultry meal," or "corn gluten meal")

➤ Menadione (synthetic form of vitamin K)

➤ Peanut hulls (a significant source of mycotoxins)

➤ Dyes and colors (for example Red #40), including caramel

➤ Poultry or animal digest

➤ Animal fat

➤ Propylene glycol

➤ Soybean oil, soy flour, ground soybeans, soybean meal, soybean hulls, or soybean mill run

➤ "Oxide" and "sulfate" forms of minerals (for example, zinc oxide, titanium dioxide, or copper sulfate)

➤ Poultry or beef by-products

➤ Butylhydroxyanisol (BHA), butylhydroxytoluene (BHT), and ethoxyquin (synthetic preservatives)

➤ Sodium selenite (synthetic form of selenium)

Evaluating Product and Process

When evaluating brands, product *and* process are important. **"Read before you feed"** is our suggestion for every product that goes into your dog's mouth. Brands differ by region and country, but how you evaluate foods remains the same, beginning with your personal food philosophy. All the information you need about a prospective food should be on the company's website. If the food is organic, made with human-edible ingredients, or GMO-free, the website will tell you all about it. If the information you're looking for is missing from the website, it's most likely missing from the product. Pet food companies use their websites to highlight their most attractive product benefits—you won't have to dig. If you have questions, email or call the company. After you've used our checklist to develop your own personal food philosophy, you'll move along to Pet Food Homework.

Each ingredient in a bag of dog food has a history—an important story to tell. The quality and amount of raw ingredients in dog food, and how each ingredient is altered or adulterated, ultimately dictate how biologically appropriate, wholesome, and healthful the food product is. Yes, it's a little cumbersome to do online research to know what's in the food your dog is eating, but it's the only way to find out—and your dog's health depends on it.

Three Clarifying Calculations to Assess Dog Food

The great news is all dog food products can be evaluated using three easy assessments that level the playing field, by eliminating confusing marketing hype. You can do some simple math—Count Your Carbs, Adulteration Math, and Synthetic Nutrition Addition—to compare dog food brands, side by side. Each calculation yields a score that can be stacked up and compared to other brands; each score falls into a good, better, or best category, compared to the competition. As you read this information, you may find yourself prioritizing the results according to what matters most and your personal food philosophy. This is exactly what we want you to do: Focus on what makes the most sense to you right now.

Count Your Carbs

Pop quiz: How many carbs does a dog require? We hope you shouted "Zero!" A dog's carb requirement is zero, but they—like us—love their carbs and can be big fat-carb addicts. Consuming 30–60 percent starch for energy (what most kibble shakes out to be) delivers results similar to what we see in our kids in their fast-food world. That much starch creates lots of energy (and those calories can lead to obesity) and poor brain chemistry, inflammation, and nutrient deficiencies (overfed and undernourished) because the carb calories displace the much-needed calories from nutrient-dense fresh meat.

Calculating the carb (starch) content in foods is a powerful tool to determine if the food is biologically appropriate. A dog's evolutionary diet was high in moisture, rich in protein and fat, and very low in sugar/starch—the exact opposite of kibble.

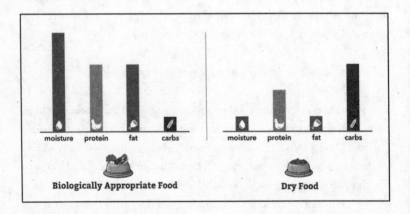

Pet food carbs (millet, quinoa, potatoes, lentils, tapioca, corn, wheat, rice, soy, chickpeas, sorghum, barley, oatmeal, "ancient grains," etc.) are far cheaper than meat of any quality and even meat by-products and meat meals, and their stickiness helps hold the food together during the manufacturing process. So the carb calculation also tells you what you're paying for: cheap, unnecessary starch or more expensive meat.

Remember, when we say carbs, we aren't talking about healthy fiber, which is sugar-free. We mean the "bad carbs"—starch that turns into sugar that creates metabolic chaos and harmful AGEs. These are the carbs we need to minimize in our dog's bowls. And these are the carbs that comprise the bulk of many (dare we say most) ultra-processed pet foods. Animal nutritionist and pet food formulator Dr. Richard Patton told us wild dogs rarely had access to foods that contained more than 10 percent starch. Because knowing dogs don't have a starch requirement, less is more.

Remember, when given a choice, dogs select protein and healthy fats, not carbs. It's not that you have to obsess about eliminating *all* starch from your dog's bowl. Just like us, dogs can consume some foods that are metabolically stressful (aka fast food), but the goal is to prioritize calories from foods that resonate with his innate metabolic machinery—lean meats and healthy fats.

By now you're aware that consumers are long past the notion of just wanting to see meat first on a dog food label. Why did the meat fail inspection and end up as a "feed ingredient"? Was it healthy "trim" meat or diseased tissue? Where did the meat come from? Savvy consumers know the tricks of the trade: ingredient splitting and the salt divider. What isn't so clear is the amount of energy, or calories, coming from cheaper starchy carbs, because at the time of this book printing, we still don't have Nutrition Facts labels on pet foods.

Feeding a diet less than 20 percent starchy carbs is the most nutritious, least metabolically stressful food for *Canis lupus familiaris*. Minimizing starch intake also minimizes your dog's consumption of toxic "cides," including herbicides, pesticides, glyphosate, and mycotoxin residues that are passed up the food chain in these crops, many of which have been genetically modified. A growing number of pet food companies now provide information about carbohydrate content, either on the product or on the

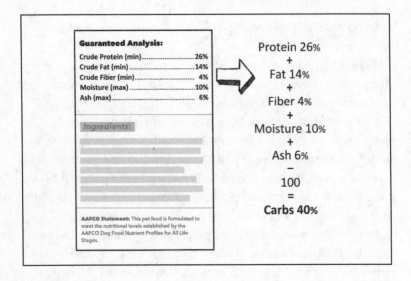

website. If this information isn't provided, call the company and ask, but it's faster to just calculate it yourself. Dry and moist foods require slightly different equations to account for the food's water content (you can find the equation for canned/moist foods at www.foreverdog.com).

To calculate the carbs in dry food, find the Guaranteed Analysis on the side of your dog food bag or on the website. The Guaranteed Analysis lists the amount of crude protein, fiber, moisture, fat, and ash in the diet. Ash content is an estimate of the mineral content in the diet. Sometimes pet food companies don't list the ash content in the Guaranteed Analysis. If you don't see the ash content, assume it's 6 percent (ash content ranges from 4–8 percent in most foods). To calculate the starch content, simply add the protein, plus the fat, plus the fiber, plus the moisture, plus the ash (6 percent if not listed) and subtract from 100. That number is the percentage of starch (aka sugar) in your dog food. We recommend sitting down while doing this equation as many people are shocked to discover their "super-premium" $120 bag of dog food is a whopping 35 percent starch (sugar).

Good: dog foods with less than 20 percent carbs from starch
Better: dog foods with less than 15 percent carbs from starch
Best: dog foods with less than 10 percent carbs from starch

On occasion there are medical reasons nutritionists or veterinarians increase a dog's carb content (during pregnancy, for instance); but generally speaking, unlike goats and rabbits, healthy dogs don't require the bulk of their calories from carbs, so we don't recommend carb-loading your dog unless it is medically necessary.

Adulteration Math

This second assignment in Pet Food Math helps you determine the level and intensity of processing. The more refined and altered the food becomes, the less nutrition and the more toxic processing by-products are present. Determining the fresh, flash-processed, processed, and ultra-processed status of foods can be difficult, but we will make it as simple as possible for you. To recap what you learned in Part II:

Unprocessed (raw) or "fresh, flash-processed foods": Fresh, raw ingredients that are slightly altered for the purpose of preservation with minimal nutrient loss. Examples include grinding, refrigeration, fermenting, freezing, dehydration, vacuum packaging, and pasteurization. These minimally processed foods have one adulteration.

Processed foods: The previous category (flash-processed foods) that are modified again by an additional heat process, so two processes.

Ultra-processed foods: Industrial creations that contain ingredients not found in home cooking; require several processing steps using multiple previously processed ingredients, with additives to enhance taste, texture, color, and flavor; and are produced through baking, smoking, canning, or extrusion. Ultra-processed food has multiple heat adulterations. The average bag of dry food contains ingredients that have undergone high-heat processing an average of four times.

Fresh, flash-processed foods have been manipulated (adulterated) fewer times and with no to low heat. Why is this critically important? The enemies of nutrition are time, heat, and oxygen (which causes oxidation, leading to rancidity). With dog food, heat is the most pervasive offender. Heat negatively impacts the level of nutrients in the food; each time the ingredients are heated, more nutrients are lost. There is no publicly available research on the extent of nutrient depletion through ultra-processing on a brand-by-brand basis. However, *lots* of synthetic vitamins and minerals are added back in to account for profound nutrient loss during processing, which provides insight into just how devoid and nutritionally depleted the end products are. We've pulled an example of nutrient losses from the human literature to demonstrate what happens to a few nutrients after *one* heat process. Look at the "reheat" values for an idea of what happens after three additional reheats in an average bag of dry dog food.

Typical Maximum Nutrient Losses (as compared to raw food)					
Vitamins	Freeze	Dry	Cook	Cook+Drain	Reheat
Vitamin A	5%	50%	25%	35%	10%
Vitamin C	30%	80%	50%	75%	50%
Thiamin	5%	30%	55%	70%	40%
Vitamin B12	0%	0%	45%	50%	45%
Folate	5%	50%	70%	75%	30%
Zinc	0%	0%	25%	25%	0%
Copper	10%	0%	40%	45%	0%

The bad news doesn't stop there. Each time ingredients are heated we lose more of our most powerful weapons against aging and disease. Potent polyphenols and enzyme cofactors that positively influence our dog's epigenome are cooked off, fragile essential fatty acids that create resilient cell membranes are inactivated, and proteins and amino acids are denatured. Repeated heating also obliterates the "entourage effect" of whole, raw foods—the diverse microbial community in each fresh food that works in harmony with the naturally occurring vitamins, minerals, and antioxidants to provide exactly what our dogs need to support thriving bodies. All of it gone.

Ultra-processed diets cause parallel damage: Repeated heating *eliminates* nutrients and bioactive compounds that prevent disease and degeneration, and it *creates* biotoxins that accelerate the cellular aging and dying process. **Advanced glycation end products**

(AGEs) created through heat processing rapidly age our dogs
and create disease—and our dogs eat massive amounts of
these toxic substances every day in ultra-processed pet foods.
Heating ingredients over and over creates microscopic monsters
that the pet food industry desperately wants to ignore. Repeated
Maillard reactions create AGEs in the end food products that negatively impact a dog's health in about every imaginable way. The
hotter the heat, the longer the food is heated, and the more times
a food is heated, the more AGEs are produced. Repeated heating
diminishes the nutrient composition and increases the amount of
AGEs in the food.

The better the quality—and the broader the spectrum—of raw
ingredients, the more initial nutrients coming from real, whole
food present in the product (especially important when thinking
about raw brands). Obviously, the fewer times the ingredients are
heat-treated, the more nutrition that remains in the end product.

How to calculate the level of heat-processing: Adding up how
many times the ingredients in your dog food have been heat-
adulterated is simple math, but determining how each type of food
is produced can be a little more difficult. Let's look at a few examples, so you can see the difference.

Dry foods: Animal carcasses are ground up and boiled to separate animal fat from bone and tissue, a process called "rendering"
(first heat adulteration). Bone and tissues are strained and pressed
to remove moisture, heat-dried (second heat), and pulverized to
make meat meals. Peas, corn, and other veggies you see listed on
the label most likely arrived at the pet food plant already dried (via
heat) or in powdered form (like pea protein and corn gluten meal).
These dry, already-heat-processed ingredients are then blended
with other ingredients (that have also been previously cooked and
dried) to make a dough that is high-pressure-cooked in an extruder, baked, or "air dried" at high temperatures. Extruded kibble
is heated a fourth time when it comes out of the extruder to reduce
the moisture content, the final step in the process (and at least a

fourth heat adulteration). **A typical bag of dry dog food has been heated and processed so many times (four!) that most nutritional value has been, quite literally, destroyed.**

On the opposite end of the spectrum, unprocessed raw diets contain fresh ingredients that have never been heated, only blended together to meet optimal nutritional parameters and be fed. If raw ingredients are blended and then undergo one flash (quick) adulteration, they're considered flash-processed. This includes:

Frozen raw dog food: Raw ingredients are blended together and frozen. Raw dog food is sterilized to remove bacteria with cold-water pressure (high-pressure pasteurization, or HPP), a second nonthermal adulteration.

Freeze-dried dog food: Fresh or frozen meat is blended with fresh or frozen vegetables, fruits, and supplements, then freeze-dried (one adulteration, no heat). If the ingredients were frozen first, two processing steps occurred—but, without heat, so nutrient loss and AGE production are negligible.

Gently cooked dog food: This is one of the fastest-growing segments of the pet food industry, and for good reason. A surge of very successful, super-transparent companies now make human-grade dog food that is semi-customizable and incredibly convenient. Most of these companies have mastered the customer experience: Their websites allow discerning pet parents to input their dog's age, weight, breed, and exercise habits, along with any food sensitivities or dietary preferences, and they then ship tailored meals or customized meal plans (frozen) directly to the customer's door, with recurring auto-shipments. Want only organic ingredients? No problem. Have a dog that's allergic to several proteins? No problem. No wonder these companies are providing good competition to the other pet food categories. The healthiest of these cooked diets extend shelf life via freezing, so you'll find these diets next to the raw and pasteurized raw (HPP) diets in the freezer section of your local pet store.

But even some very popular cooked, refrigerated dog food brands falter when it comes to ingredient sourcing transparency and

Synthetic Nutrition Addition. Questions like "Where does your meat come from? How can the meat last *six months* in the refrigerator?! How is it preserved?" can result in long, awkward pauses when you call customer service. A long list of synthetic vitamins and minerals gives rise to questions about the nutritional density of the companies' raw materials and heat-processing techniques. This may be no big deal to you or a very big deal, depending on your personal food philosophy; but these are the types of commonsense questions we encourage pet parents to ask about whatever brand they're feeding to their animals.

Dehydrated dog food: Many brands pass with flying colors. They contain minimal starch, start with all raw ingredients, and are dehydrated at low temperatures for shorter periods of time. And several dehydrated dog food brands achieve *no* "good" marks. Moral of the story: investigate the product label to learn more. Count Your Carbs to see if the calories come from real meat and healthy fats or starch. Dog food companies starting with raw, fresh ingredients will list them as such (for example, chicken, green beans) on the label. If the ingredients listed are "dehydrated chicken, dehydrated green beans" the ingredients were shelf stable (not fresh) before they became dog food, so at least one additional heat step occurred at the ingredient supplier. Lastly, Synthetic Nutrition Addition (the next equation we'll teach you) will help you decide if the dehydrated brand you're considering resonates with your personal food philosophy.

Look at the ingredient label to get an idea of how many times the ingredients have been processed using heat. If the food must be kept frozen (is not shelf stable), that's your best indication of freshness. If the food is shelf stable (freezing not required), some process was used to stabilize the food. Freeze-drying is the least harmful to nutrients, followed by low-temp dehydration. If you have questions about whether the ingredients are fresh or preprocessed (dried), call the company and ask. The lower the adulteration score, the healthier the food.

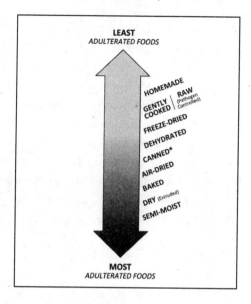

We interviewed AGE expert Dr. David Turner, PhD, from the Medical University of South Carolina. He explained the latest research comparing dog food processing techniques and AGE production showed canned food (cooked at 254 degrees F) had the highest levels of AGEs. This was potentially due to the amount of sugar/starch in the food, the cumulative effects of AGEs in ingredients used, and the longer duration of time canned foods are heated. Conversely, other studies note that higher moisture foods (such as canned products) may buffer AGE production, so depending on the starch content, temperature, and the time canned food is heated, there can be a wide range of AGE levels, hence our asterisk by canned food. Semimoist food gets an F across all Pet Food Homework assignments because there's no good/better/best options in this category. We simply don't recommend feeding semimoist food, ever.

In some cases, it is hard to know what processing technique was used to make the food. These days, manufacturers go to great lengths to avoid calling their products anything that resembles traditional kibble, including creating their own descriptions such as "clusters," "chunks," and "morsels." The new category of extra-confusing dry foods includes "raw coated" kibbles: AGE-loaded pellets that have a layer of freeze-dried raw food on the outside, to make it sound healthier. Like adding a pinch of broccoli sprouts to your Big Mac and fries, the good doesn't balance the bad with these expensive fast foods.

The term "minimally processed" is the new industry buzzword, and companies across all pet food categories use the term in their marketing material, regardless of their processing technique. Despite suggestions that the pet food industry adopt guidelines defining "minimally processed," nothing has been officially adopted, so the term deceptively encompasses all processing techniques, excluding extrusion. This is why we recommend relying more on Pet Food Homework results, and not company marketing claims, for the most unbiased and eye-opening food evaluation process. If you can't tell how the food was processed by reading the company website, email or call and ask how many times the ingredients in the food have been heated, to what temperature, and for how long.

We found it interesting to learn that even raw chicken meat may contain low levels of AGEs if the chickens were factory farmed and consumed high-heat-processed chicken feed (full of glyphosate and AGEs) prior to slaughter. AGEs are passed up the food chain. In the pet food AGE research study, extruded dog food (cooked at 265 degrees F) is the second-worst offender, in terms of harmful AGE levels. Of course, raw food has the least amount. Similar to canned foods, "air-dried" foods are all over the map because of starch content and wide temperature variations, so it really does pay to call the company and have a chat (unless you can get all the info you need from the company's website).

Adulteration Math Results

Good: *previously processed* ingredients blended together, and *heat processed once* (many dehydrated foods).

Better: *raw, fresh* ingredients blended together and freeze-dried or high-pressure pasteurized (HPP), as well as raw fresh ingredients blended together and *no-heat or low-heat processed* once (many raw meat dehydrated foods and gently cooked foods).

Best: *raw, fresh ingredients* blended together and served, or frozen (*no heat process*) to be eaten within three months (homemade food, commercial frozen raw food).

Synthetic Nutrition Addition

The last assignment in your Pet Food Homework allows you to determine the *source* of nutrition in the food. To recap, the number of added vitamins and minerals in the product reflects the deficiencies of vitamins and minerals from the raw ingredients and/or added to make up for the lost nutrition burned off during the intense heat processing that inactivates the once-present nutrients. Required nutrients come from one of two sources: nutrient-dense food ingredients or synthetics (added lab-made vitamins, minerals, amino acids, and fatty acids). The less nutrient-dense the dog food and/or the more heat used to produce the food, the more synthetics that must be added.

The good/better/best scoring on this assessment is the most subjective of the three assignments we provide, depending on your personal food philosophy. In our experience, dog owners tend to feel strongly about this subject, one way or the other. The folks who don't really have an opinion point out that we all eat synthetic vitamins and minerals in many of the fortified foods we consume; other pet parents take lots of synthetic vitamins and mineral supplements themselves. For these folks, it's more acceptable for their dogs to derive the bulk of their micronutrients in the same way. The beauty of Pet Food Math is that you get to decide what's right for you and your pooch, in light of your personal food philosophy. The math is simply a tool to allow you to make informed decisions about your dog's diet and health.

Adulteration Math is about how many synthetic nutrients must be added back into the product to make it nutritionally adequate. Poorer-quality ingredients (usually a food versus feed issue) and fewer nutrient-dense ingredients (always a cost issue) necessarily mean more synthetics. In addition to counting the number of synthetic vitamins and minerals added, also scan the label for added nasties, aka the Dirty Dozen: ethoxyquin, menadione, dyes and colors (including caramel), poultry (animal) digest, animal fat, propylene glycol, soybean oil, by-products, corn gluten meal, BHA/BHT, meat meals, and sodium selenite.

How to do it: Count the number of synthetic nutrients on the food label (you can find the ingredient list on the company's website or on the back of the package). While you're surfing the company's website, keep your most important food philosophy points in mind. Added vitamins and minerals can be found on the ingredient panel after the food ingredients (see the diagram). Each nutrient is separated by a comma, so even though you may not be able to pronounce them, you can count how many are included.

Good: Dog food that contains none of the Dirty Dozen on the label (found on page 288) and *fewer than twelve* synthetic nutrients.
Better: Dog food that contains none of the Dirty Dozen on the label and *fewer than eight* synthetic nutrients, with some added health perks: organic ingredients, some GMO-free ingredients, etc.
Best: Dog food that contains none of the Dirty Dozen, and *fewer than four* synthetic nutrients, with lots of added health perks:

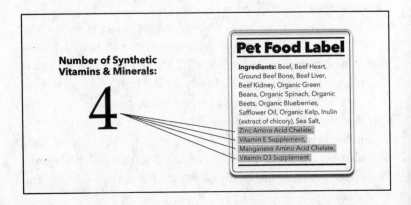

Number of Synthetic Vitamins & Minerals:

4

Pet Food Label

Ingredients: Beef, Beef Heart, Ground Beef Bone, Beef Liver, Beef Kidney, Organic Green Beans, Organic Spinach, Organic Beets, Organic Blueberries, Safflower Oil, Organic Kelp, Inulin (extract of chicory), Sea Salt, Zinc Amino Acid Chelate, Vitamin E Supplement, Manganese Amino Acid Chelate, Vitamin D3 Supplement human-grade ingredients, organic, GMO-free, wild-caught/free-range/pastured meats, etc. These are the most expensive products in each food category, because the nutrients come from expensive, real-food ingredients and not a vitamin-mineral premix.

This exercise is intended to provide you significant discretion to determine what's right for you and your pooch, based on your values, beliefs, priorities, and budget; and there are many variables to consider. For instance, raw diets haven't been heat-processed (so there's no loss of nutrients via heat and no AGEs created). If you see a wad of synthetic nutrients on a nutritionally complete and balanced raw food label, it means the company is leveraging synthetic add-ins to provide the minimum nutrients your dog needs (so you'll see a less diversified ingredient label, maybe just meat and organs, plus the list of added synthetics). This food will be cheaper because the company isn't buying expensive or exotic fresh ingredients to meet specific amounts of missing nutrients. Compare that with a raw food that has two synthetics (usually vitamins E and D). These labels contain a long list of expensive food ingredients, because that's the source of the nutrition.

For those of you who decide to continue feeding kibble (plus 10 percent CLTs), how do you evaluate kibble brands?

Kibble should be evaluated the same way we evaluate fresher food brands. Pet Food Homework (calculating carbs, adulteration math, and counting the number of synthetics) and ranking results according to the good/better/best system can be used for any type of dog food. It is particularly important to watch out for the Dirty Dozen. There is a big range of quality and varied processing techniques within the category of kibble: "Cold extrusion," "gently baked," and "air-dried" are newer thermal processes completed at varying oven temperatures; the starch content of these products also varies widely. Your personal food philosophy comes into play with every product you buy for your dog, so ask the same questions about the bag of kibble as you would any other food. When it comes to cost, compare categories wisely, especially with the more expensive dry food brands. If cost is an issue for you, keep in mind an organic "super-premium" kibble can be more expensive than frozen fresh food, delivered right to your door. It pays to do your research.

Pro Tip: Kibble is more prone to rancidity than other types of dog food, so it's imperative to store kibble in a cool, dry place (ideally the freezer) and buy small bags you can feed within three months, and ideally within thirty days.

THE GROWING POPULARITY OF RAW DIETS

The 2.0 pet parents are flocking to raw pet foods; minimizing starch, AGEs, and synthetics translates into rock-star grades in the Pet Food Homework exercises for raw dog food brands. Delicate, heat-sensitive food enzymes, essential fatty acids, and phytonutrients remain intact and ready to be passed up the food chain and into your dog's body. Approximately 40 percent of commercial raw dog foods in the United States have undergone nonthermal high-pressure pasteurization, one of several FDA-accepted processes companies use to adhere to the zero tolerance policy for salmonella in pet foods. Make sure the raw diets you choose are clearly labeled with a statement of nutritional adequacy, because it's the biggest problem with this category.

Your Pet Food Homework provides a frame of reference for you when evaluating brands. Most important, they're assessment tools to help you better understand more about the brands you want to feed (or avoid). Don't worry; there is no right or wrong here. All that matters is that you can make smart, educated choices backed by science to fit your lifestyle and best meet your dog's needs. It's important to distinguish between what's ideal and what's realistic when it comes to your Forever Dog Meal Plan. Rarely can any of us do everything right all the time; it's about starting somewhere and feeling good about the incremental changes we make that positively influence our dogs' health. Sometimes gaining knowledge results in feeling guilty, and the more we learn, the more inadequate we feel. Shifting our mindset from overwhelmed to empowered is a good first step.

Of course, there are all sorts of caveats within the good/better/ best system, so a little common sense and discernment will serve you well. Pet Food Math is best used with brands that claim to be nutritionally complete. For instance, if you decide to try a dog food labeled for "supplemental or intermittent feeding" (a nutritionally deficient diet you will balance yourself), you'll find that nutritionally inadequate brands can fall into the "best" category because synthetic vitamins and minerals aren't added. This certainly does not make them the "best" (unless you fix the deficiencies yourself). We recently saw a pet food at a local farmers' market labeled something like this: "free-range duck meat, duck heart, duck liver, organic spinach, organic blueberries, organic turmeric." That's a lovely start, a good base, but there's no source of iodine on the label for healthy thyroid function, among a wad of other missing sources of vitamins and minerals. You may not have enough nutrition knowledge to look at a label and know the diet is iodine deficient now—or maybe ever—but you can learn to ask good questions.

The great thing about the booming dog food industry is there are so many options to choose from, and more brands enter the market almost weekly. We recommend finding several companies you like and rotating between different brands and proteins. Rotating brands is one of the best ways to provide nutritional diversity when feeding commercial dog food. Initially, this may seem overwhelming, but in time you'll come to appreciate all of the feeding options and styles available, as there's something for everyone (and every dog), depending on your food philosophy, time, and budget. You can mix and match foods with different good/better/best scores; rotate brands, proteins, and categories of food; and supply a virtually endless option of recipes and feeding styles to customize the perfect longevity bowl for your dog. And the next bag you buy or batch you make can be completely different. But before you make a purchase, do your homework so you know what you're buying.

Exercise 2:
Decide on a Type of Fresher, Flash-Processed Food

There are no hard-and-fast rules that apply to all dog foods because there are so many variables to consider. Only you can assess how these variables play into your lifestyle and your dog's unique needs. If you're new to the fresh-feeding community, deciding what type of fresher food to feed can be the most confusing and daunting aspect of turning over a new food leaf.

Within the fresher category of pet foods there are many different types of diets, including homemade diets and commercial raw, cooked, freeze-dried, and dehydrated dog food diets that you can purchase locally (independent pet food retail shops are a great place to start). You also can order all types of fresher foods online from shockingly remote places with great success. These very diverse, fresher diets create the need to make more decisions. There are many aspects to consider, all pertaining to your life circumstances and personal food philosophy. We highlight some of the benefits and shortcomings of each type of diet and provide clarity on a few confusing topics you may encounter in your search. Our goal is to provide an overview of all your fresher food options so you can then decide what recipe or company will best meet your needs. Next, we cover all the fresher food options available to you. As you read about each pet food category, think about which categories make the most sense for your lifestyle, your budget, and your dog.

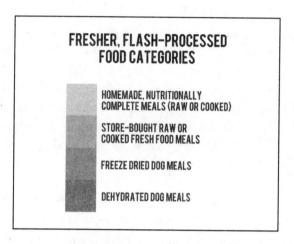

Homemade, Store-Bought, or Hybrid Meal Plans

Homemade

A homemade meal certainly offers you the most control over the ingredients in your dog's food, but making homemade meals for your pooch can be both expensive and time-consuming. You'll also need freezer space, unless you plan to prepare food every day, which is wonderful but can quickly become overwhelming. Most of us make homemade dog food once a week, once a month, or even every three months, freezing meals in small batches for easy thawing. Vets often warn against homemade meals because people do it incorrectly—meaning they *guess* at how to provide the minimum nutrient requirements to their dog. Here's the super-abbreviated backstory on "balanced diets" (a term that doesn't really mean anything because everyone's definition is different).

As we've already explained, the National Research Council (NRC) developed minimum nutrient requirements, the basic amounts of vitamins and minerals that puppies and kittens, pregnant and lactating females, and adult dogs need to not develop nutritional deficiencies. These experiments were conducted years ago, and not in the most ethical way: The researchers denied the test animals each nutrient and recorded what happened (or didn't happen) clinically, then sacrificed the animals, recording what happened or didn't happen, internally. We know conclusively—there is *no* debate about—the *minimum* levels of micronutrients needed to prevent a myriad of nutritionally related diseases; we also learned, in some cases, what the consequences are of supplying nutrients in too high quantities or in the wrong ratios. After the NRC published its minimum nutrient requirements, AAFCO (US) and FEDIAF (Europe) used the NRC information as a basis for creating their own standards. Many critics argue that all nutrition standards are flawed, and we agree. But we can also agree that nobody wants to unwittingly conduct nutritional deficiency experiments in their own kitchen, with their personal pet. Research provided clear instructions on the milligrams of each vitamin, mineral, and fatty acid a dog needs, per 1,000 calories of energy consumed, to sustain life. The problem is twofold: Merely sustaining life isn't our goal, *we're Longevity Junkies!* Second, guessing how to provide those basic, required nutrients (and in what quantity) is hard, and most people guess incorrectly (thus vets caution against homemade diets). We wrote this book to help our fellow dog lovers understand why feeding *nutrient-dense* diets matter, and how to do it right.

Some raw dog food proponents argue NRC developed its minimum nutrient requirements using research on animals fed ultra-processed foods, not the unadulterated meats dogs and cats evolved consuming. We agree; that certainly could skew results. Until researchers collect sufficient data from animals fed raw and flash-processed foods, we're left with current standards in assessing whether we're in the ballpark of our dog's basic nutritional requirements. But there is good news: Research shows fresher foods are more digestible and assimilative. When we assess raw and flash-processed diets using current minimum nutrient standards (a very low bar), it's a home run: fresh food delivers *optimal* levels of whole-food nutrients. Said another way: If your raw or flash-processed dog food recipe meets the current less-than-ideal standards it will provide *superior nutrition* compared to other types of ultra-processed diets.

The problem: many well-meaning guardians assume that rotating a variety of fresh meats, organs, and veggies over time will give their dog everything they need nutritionally. The end result can be devastating. We've seen many heartbroken clients grieving. We've heard (loudly) vets and veterinary organizations saying, "Don't feed homemade diets, they're very risky!" Vets have seen enough pain and misery among well-intentioned pet owners and their dogs to justify their concerns about homemade food. In short, we don't blame vets for their skepticism: homemade diets can be the best or worst foods, for this reason.

We have a solution that will simplify your decision making and give you the confidence you need to make important changes that can transform your dog's health. The solution: Prove to your vet that you can do this right by following a recipe or template that has been formulated to meet the known nutrient requirements of dogs. Explain to your veterinarian that you are learning more about pet nutrition and you plan to follow guidelines that prevent nutrient deficiencies from occurring. Frankly, this will ease the collective anxiety in the exam room.

If you choose to make dog food at home, we applaud you! We encourage you to pause for a moment and appreciate your life's blessings—you *have* this option; many do not. We admire your dedication to optimizing your pup's health and longevity. This is a commitment that pays herculean dividends!

We also implore you to follow a variety of recipes or use nutritional assessment tools that ensure your meals at least meet the bare-bones minimum nutrient requirements for vitamins, minerals, amino acids, and essential fatty acids (see an example of what you should look for when evaluating homemade recipes on our website). You can serve homemade food to your dog uncooked (raw) or cooked (we recommend poaching, which creates the least amount of AGEs, but cook it however you wish). The vast majority of people spending the time, energy, and resources to make homemade diets recognize that *nutritionally optimal* diets are very different from *minimally adequate* diets. These Longevity Junkies truly understand the power of whole-food nutrition and want to maximize this powerful tool for their dogs' benefit.

What are synthetic nutrients? Laboratory-made vitamins and minerals are synthetic nutrients—they are man-made and used in both human and animal food products to fortify diets. There is a wide range of synthetic nutrients that vary in form and type (which dictates digestibility, absorbability, and safety), as well as quality and purity. The more vitamins and minerals you and your dog get from eating real food, the fewer synthetic nutrients you will need.

There are two categories of homemade dog food: (1) whole-food recipes (synthetic-free), and (2) recipes with synthetic nutrients.

Synthetic-Free Homemade

In the homemade, whole-food recipes (synthetic-free) category, all of the nutrients are supplied by whole foods. No additional vitamin or mineral supplements need to be purchased to meet your dog's nutritional needs. Sometimes hard-to-come-by and more expensive ingredients are needed, like Brazil nuts for selenium, and canned oysters or clams for zinc. When specific foods supply vitamins and minerals to your dog's body, his body knows exactly what to do with them because *they come from real food.* But you need to follow a whole-food recipe exactly—we mean precisely—to meet minimum nutrient requirements. In the wild, dogs eat a range of different prey species and a wider array of body parts (including the eyes, brains, and glands) to obtain all required vitamins and minerals. Take zinc: We don't know many people who feed their dogs testicles, teeth, and rodent hair (great food-based sources of zinc), so zinc can be a scarce nutrient; simply rotating through traditional cuts of meat from the grocery store and a variety of veggies won't meet your dog's minimum zinc requirements. Zinc deficiency leads to poor skin health, poor wound healing, and GI, heart, and vision problems. This is true for other hard-to-come-by nutrients as well, including vitamins D and E, iodine, manganese, and selenium, to name a few.

Unfortunately, all-in-one dog multivitamins aren't sufficient to balance homemade meals. "Dietary drift" also occurs when people start to swap out ingredients and the recipe becomes unbalanced. Drift sets the stage for nutritional problems, as does rotating through a variety of nutritionally incomplete recipes. This also rightfully agitates your veterinarian.

Homemade meals can be fed raw or gently cooked (poached, simmered, stewed). When handling or feeding raw meat, you must follow safe food-handling techniques for preparation and storage, just as you would do for yourself. The same foodborne risks apply to all meat, whether it's destined for your barbecue and your belly or your dog's bowl. Healthy dogs are evolutionarily adapted to deal with a much heavier bacterial load, and their hyper-acidic stomach expertly manages incoming microbes. E. coli, salmonella, and clostridium are all found in the GI tracts of healthy dogs, even those eating kibble; they are "normal inhabitants."

HOW TO POACH FOODS

Poaching gently cooks while preserving nutrients and moisture. Foods do not brown when they are poached, so fewer MRPs are produced. Place meat in a pot, add filtered water (or Longevity Junkie Homemade Bone Broth, page 242, or Medicinal Mushroom Broth, page 225) to just cover the food in liquid. Cooking experts say to add a splash of raw apple cider vinegar to "set" the proteins; we don't have any science to document this step—we do it because the pros say it's a good idea. Heat to 160 degrees F (70 degrees C), which kills bacteria but does not create massive AGEs. Cook times vary, depending on the amount of meat you're cooking (usually five to eight minutes for small batches). Save the remaining nutrient-dense liquid to top off meals just before feeding. You can also add in herbs and spices (see page 225 to create more polyphenol-rich and flavorful broths).

Homemade whole-food diets are the most expensive recipes you can feed (especially if you choose organic, free-range ingredients), but they are also the most nutritious and freshest food you can feed. Conventional produce and factory-farmed meats can reduce the cost. That said, wild-caught, pastured, free-range meats can be more nutrient dense and have lower chemical loads. We recommend supporting your local farmers. In cities, check out your local farmers' market or food co-op to see where to find locally grown produce and meats. Independently owned health food stores often can point you in the right direction of locally sourced meats and produce. You can custom curate rare meats if your dog has allergies; you can add amazing superfoods with specific health benefits for your dog's medical or nutritional needs. Most important, you know exactly what your dog is eating because you have hand-selected all of it yourself. Many of the professionals listed in the www.freshfoodconsultants.org directory offer nutritionally complete, ready-to-download homemade dog food recipes.

If your dog requires a specific "therapeutic" diet for medical issues, there are board-certified veterinary nutritionists, worldwide, who can formulate a homemade recipe specifically for your dog. Find one at www.acvn.org. The veterinary nutritionists at www.petdiets.com will formulate custom raw or cooked recipes for dogs with medical issues or specific health goals.

When you Google "homemade dog food recipes," you'll find endless links to sites that feature beautifully displayed bowls that resemble those found on the Food Network for human consumption. Again: be very careful. Homemade recipes (whether you find one online or in a book) should be clearly labeled with a nutritional adequacy statement: "This recipe was formulated to meet minimal nutrient requirements according to _____ standard" (AAFCO, NRC, or FEDIAF). The recipe should also come with a list of ingredients by weight or volume, specify the leanness of meat required, list the calorie content, and provide a breakdown of the amounts of vitamins, minerals, amino acids, and fats in the recipe (see example on page 302). Don't use recipes that do not provide this information except as treats or toppers (up to 10 percent of your dog's calories) or an occasional meal. Relying on unformulated recipes as your dog's base diet can result in nutritional deficiencies that negatively affect health span. We provide more examples of nutritionally complete recipes at www.foreverdog.com; below is one such example. Most people just getting started find it much easier and more convenient to freshen up their bowls with commercially available, well-formulated frozen diets from trusted companies that follow AAFCO, NRC, or FEDIAF nutritional guidelines. However, if you love to cook or prepare food, you'll make your dog very happy!

This is an example of a nutritionally complete meal for adult dogs *using whole foods* (puppies need more minerals than this recipe supplies; puppy recipes are far more complex). Note that you must use extra-lean (90 percent plus) ground beef and add key foods to meet specific nutrients. For instance, the raw ground sunflower seeds provide the vitamin E, the hemp seeds provide the required alpha-linolenic acid (ALA) and magnesium, cod liver oil provides vitamin A and the 1,300 IU of vitamin D needed, ginger from your

HOMEMADE BEEF DINNER
FOR ADULT DOGS

5 pounds	(2.27 kg)	Extra lean ground beef, poached or raw
2 pounds	(900g)	Beef liver, poached or raw
1 pound	(454g)	Asparagus, finely chopped
4 ounces	(114g)	Spinach, finely chopped
2 ounces	(57g)	Raw sunflower seeds, ground
2 ounces	(57g)	Raw hempseeds, shelled
	(25g)	Calcium carbonate (from your local healthfood store)
	(15g)	Cod Liver Oil
	(5g)	Ground ginger powder
	(5g)	Kelp powder

spice drawer provides the manganese, and iodine-rich kelp provides the mineral needed for adequate thyroid function. If any one of these ingredients isn't provided in the amounts specified, the recipe becomes unbalanced—fine for a treat or a one-off meal, but inadequate as the staple food for your dog. Most important, the appearance of well-rounded recipes must be validated by a nutritional analysis, to confirm you're actually in the ballpark of meeting your dog's daily nutritional requirements if you feed the recipe long-term.

When you look at this recipe broken down by nutrients (see page 401 of the appendix) it suddenly looks quite different—the numbers and format can be intimidating and complex. It's important that you follow feeding guidelines that assure minimum nutritional adequacy if you feed homemade meals as your dog's primary food source.

Homemade with Synthetics

The homemade with synthetics recipes use laboratory-made vitamins and minerals and other nutritional supplements to meet some of your dog's nutritional requirements. For example, instead of Brazil nuts for selenium, you will add powdered selenium from your human health food store. Like all supplements, there's a huge range of quality and different forms of nutrients you can buy, which can be empowering or daunting, depending on your knowledge and personal food philosophy.

Synthetics can be further subdivided into two distinct categories: DIY vitamin/mineral mixes, and commercial all-in-one products that are specifically designed to make homemade recipes nutritionally complete (they are not the same as general multivitamins).

DIY: Many homemade dog food recipes require you to buy individual vitamins and minerals (e.g., zinc, calcium, vitamins E and D, selenium, manganese, etc.) and add them in specific amounts to a recipe. The number of supplements and quantity added depend on the whole foods in the recipe that provide sources of nutrients; whatever *isn't* in the food must come from synthetics. Downsides to DIY blends: It can be overwhelming to buy up to a dozen individual vitamins and minerals. Grinding pills into powders or opening capsules for the correct, often minute amounts is challenging and requires precision, not to mention that the nutrients must be physically very well blended into the batch of food. Human error is a real potential. Here's an example of a homemade diet using DIY supplements (see the appendix for nutritional information). Note that including beef liver prevents the need to supplement copper and iron in this recipe.

Benefits to DIY blends: You can choose recipes that use your preferred form of supplements. Let's say your dog is prone to developing urinary oxalate crystals, and you've learned in your research that calcium citrate is the preferred form of dietary calcium for your dog's condition. You can use homemade recipes that incorporate the

HOMEMADE TURKEY DINNER
FOR ADULT DOGS WITH DIY SUPPLEMENTS

5 pounds	(2270g)	85% Lean ground turkey, raw or cooked
2 pounds	(908g)	Beef liver, raw or poached
1 pound	(454g)	Brussel Sprouts, finely chopped
1 pound	(454g)	Green Beans, finely chopped
8 ounces	(227g)	Endive, finely chopped

Supplements to add in from the health food store:

1.8 ounces	(50g)	Salmon Oil
	(25g)	Calcium carbonate
	(1200IU)	Vitamin D supplement
	(200IU)	Vitamin E supplement
	(2500mg)	Potassium supplement
	(600mg)	Magnesium citrate supplement
	(10mg)	Manganese supplement
	(120mg)	Zinc supplement
	(2520mcg)	Iodine supplement

best forms of nutrients to support your dog's specific needs. Choosing chelated forms of minerals is an option, if that's important to you. This empowerment excites some people and terrifies others. If you are looking to create nutritionally complete homemade recipes yourself, where all of the supplement math has been done for you on a spreadsheet, a subscription to www.animaldietformulator.com allows you to formulate homemade diet recipes that meet American (AAFCO) or European (FEDIAF) nutrition standards.

All-in-one dog vitamin/mineral powders that claim to balance homemade diets also come with various pluses and minuses. The biggest minus: Most are not formulated correctly to actually balance homemade diets. The majority of multivitamin and mineral products have not undergone nutritional analyses to assure nutritional adequacy for a myriad of different recipes; this can result in nutritional deficiencies or excesses over time. All-in-one products that don't meet minimum nutritional requirements or exceed safe limits lead to serious nutritional problems (e.g., bladder stones; heart, liver, or kidney disease; hypothyroidism; and growth and development issues).

We're instantly leery of the rainbow spectrum of supplements marketed with taglines like this: "Add a teaspoon of this to your dog's homemade meals to make sure he's getting all he needs."

Doctors and vets don't always correlate the medical problems we see in the exam room with nutritional deficiencies or excesses, but there can be a direct correlation. The plus of a *well-formulated*, all-in-one product specifically created to complete homemade meals: one container and no math! Add the amount indicated in the recipe to complete your homemade meal, mix well, and serve. All-in-one vitamin/mineral products are easier than blending your nutrients and decreases the risk of user error.

In general, **homemade diets with appropriate levels of synthetic vitamins and minerals are the least expensive way to feed a fresh homemade diet, and the least objectionable to vets**. The convenience of not having to source a wide variety of whole-food ingredients to meet micronutrient requirements means your dog's vitamins and minerals are coming from powders rather than whole foods; this can be a plus or a minus, depending on your food philosophy. If you use all-in-one powders, we recommend you rotate homemade recipes often to maximize the nutrient diversity coming from the fresh foods you're serving. Two well-researched options that are popular with many home-preppers: www.mealmixfordogs.com offers a complete, all-in-one powder for raw or cooked homemade meals for adult dogs; and www.balanceit.com has complete, all-in-one powders designed for all life stages (including puppies) and also has a vitamin/mineral mix specifically designed for dogs with kidney problems.

DIY HOMEMADE DIET SUPPORT

Choose ready-to-download nutritionally complete recipes:

➤ www.foreverdog.com (free!)

➤ www.planetpaws.ca

➤ www.animaldietformulator.com (its app helps you easily construct your own meals)

➤ www.freshfoodconsultants.org (links to many professionals and websites that offer ready-to-print nutritionally complete meals)

Design your own meals (choose your ingredients) with an all-in-one supplement powder:

➤ www.balanceit.com

➤ www.mealmixfordogs.com

Work with a veterinary nutritionist to formulate custom cooked diets around your dog's specific medical issues or health concerns:

➤ www.acvn.org

➤ www.petdiets.com

Work with a fresh-food consultant to create custom raw or cooked nutritionally complete recipes for your pets:

➤ www.freshfoodconsultants.org

Buy dog food formulation software and do the whole thing yourself:

➤ www.animaldietformulator.com (AAFCO and FEDIAF nutrition guidelines)

➤ www.petdietdesiger.com (NRC nutrition guidelines)

Store-Bought (Commercial) Fresher Food Diets

If you don't have the time or desire to make your own dog food, consider commercially prepared fresher dog food diets you can buy at your local independent pet store (or have delivered directly to your door). There are lots of options to choose from, all with advantages and downsides. It's worth mentioning again that all raw diets you buy should be clearly labeled with a nutritional adequacy statement; this is especially important with commercial raw food products because so many raw pet foods sold in other countries are not formulated to meet minimum nutritional requirements. In the United States, all commercial pet food must state if the diet is nutritionally complete or incomplete. Nutritionally deficient diets should be labeled "for intermittent or supplemental feeding only," meaning you can use them as treats, toppers, an occasional (once a week) meal, or, if you're a 3.0 pet parent (you're willing to invest the time and attention in an advanced approach), you can do the math and add the missing nutrients to make it a complete and balanced diet. These products are far less expensive than nutritionally complete products, so lots of people in our community choose this option, and there are entire websites devoted to helping people balance meat, bone, and organ mixes and meat grinds at home. If you go in this direction, you'll be doing a lot of math (or spreadsheet work). A number of commercial raw companies selling deficient "prey model diets" craftily employ misleading nutritional verbiage like "meets a dog's evolutionary requirements for all vitamins and minerals." If the package says nothing about the product's nutritional compliance with NRC, AAFCO, or FEDIAF, use these foods as treats or toppers only; don't feed these foods as your dog's primary food source (unless you correct the deficiencies yourself). There are plenty of well-formulated raw diets to choose from, just read the labels well.

Nutritionally Complete Raw or Gently Cooked Dog Food
(with or without Synthetics)

These nutritionally complete cooked or raw frozen meals are easy to serve; all you have to do is thaw and feed. Freezer space can be a problem, and you have to remember to thaw food for the next day. But you must trust the company and do your research. Many emerging raw food companies, especially outside the United States, are producing foods that do not meet minimum nutrient requirements, and some are made with poor-quality ingredients. One of the biggest problems in the otherwise wonderful worldwide trend toward feeding raw food is the explosion of people feeding *nutritionally incomplete* diets (or companies not following any nutritional guidelines at all). See our discussion about diets with no nutritional adequacy statement or foods labeled "for intermittent or supplemental feeding only" on page 408 in the appendix for more info.

As a reminder, there's an FDA zero-tolerance policy for potentially pathogenic bacteria in all commercially sold pet foods in the United States. The company website will provide information on how it addresses food safety with its products.

Gently cooked dog food diets also range in quality from amazing to appalling. In a nutshell, raw or gently cooked store-bought foods can be some of the best or worst foods you can buy, based on nutritional adequacy, the quality of raw materials, and company quality control. Some of the refrigerated feed-grade products available at your local supermarket or big box stores have a labeled refrigerator shelf life of *six months*, which just isn't feasible in our minds. Common sense tells us refrigerated meats should be used up in one week, max. The least-preserved, better-quality products we've found are in the freezer section. Pet Food Math is an important tool to discern good from bad in this category.

Freeze-Dried Dog Food

This can be the most expensive food on the market, ounce per ounce, because the technology and cost to freeze-dry food is expensive, but it's a great choice if you're looking for a minimally processed, shelf-stable food. It's basically raw food that's been flash frozen in a vacuum. The process of freeze-drying involves the product being frozen, the pressure lowered, and the ice is removed by a process called sublimation (the process whereby a substance such as ice goes from the state of solid to a gas, essentially skipping the state of a liquid), removing nearly all moisture.

As we've mentioned, freeze-dried dog food should be rehydrated with water, broth, or cooled tea (see page 239 for suggestions) prior to feeding (not a big deal, but another step besides scooping out of the bag). Freeze-dried food is incredibly shelf stable, which makes it great for people (and dogs) on the go. You don't need freezer space, and you don't need to remember to take food out of the freezer the day before. Some freeze-dried products are labeled as "toppers" and are not nutritionally complete: Look for the nutritional adequacy statement if you regularly use these products as your dog's entire meal.

Dehydrated Dog Food

We recommend doing Pet Food Homework prior to making any brand decisions, but doing your homework with dehydrated dog food is especially important—this category requires the most research (which is why we put it last on our list). There are two ways to make dehydrated dog foods. First, companies that make raw pet foods simply dehydrate their current raw products; these varieties are terrific, because they begin with all raw ingredients and don't contain grains or a lot of starch. These foods are fantastic, as good as freeze-dried, in our opinion.

The confusion lies in the second way to make dehydrated dog

food: Companies buy already-dehydrated raw materials, including lots of starchy carbs, then reprocess them into a dog food formula. Many dehydrated diets on the market are starch-heavy, and because ingredient suppliers dehydrate ingredients at very different temperatures (impacting nutrient and AGE load), some dehydrated foods on the market don't fall into the best (flash-processed) category. The good news is many brands do, but you really need to investigate the label.

With dehydrated diets, moisture in the food is removed slowly by low, gentle heat. Some companies producing "air-dried" dog food insist dehydration and air drying are the same processing technique, and while it's true in principle (both techniques use air to remove moisture), air drying consistently uses higher temperatures, and that's when MRPs are produced. A quick email to the company inquiring about processing temperatures will resolve any confusion. Go for brands with acceptable Pet Food Math scores that dehydrate their food at the lowest temperatures. Rehydrate these foods, too . . . no mammal was meant to eat moisture-deficient food their whole life.

Exercise 3:
Choose Your Fresh Percent: 25 Percent, 50 Percent, or 100 Percent Upgraded Bowls

It's time to start thinking about setting your first food goal: How much fresher, flash-processed dog food do you want to work toward feeding at each meal or at least several times a week? If you have no idea at this point, then think about the amount of ultra-processed pet food you'd like to reduce or eliminate from your dog's diet. To keep things simple, we preselected some basic bowl upgrades: a quarter, a half, and a full food swap to fresher diets. To take health to the next level, you can exchange 25 percent, 50 percent, or 100 percent of your dog's ultra-processed diet with fresher, flash-processed diets. Regardless of the base diet you choose to feed, 10 percent CLTs remains the same. Last but not least, if you choose not to change your base dog food at this moment, that's okay, just keep reading:

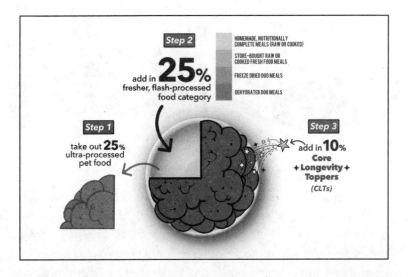

Perk Up Ultra-Processed: 25 Percent Fresher Food Really Adds Up: Swapping 25 percent of your dog's daily caloric intake with brands or diets from the fresher, flash-processed category has great health benefits. And if you think about it, adding 10 percent Core Longevity Toppers with a 25 percent fresher food upgrade boosts your dog's fresh food quotient to about one-third of her daily calories. This means you substituted about one-third of the ultra-processed food for calories coming from fresher foods—enough to make a visible difference!

50 Percent Bowl Upgrade: Replacing 50 percent of your dog's daily calories coming from ultra-processed food with calories from fresher, flash-processed dog food categories (plus up to 10 percent CLTs) means *almost* two-thirds of your dog's calorie intake is looking *very fresh*! With the 50 percent plan, your dog's daily caloric intake (about two-thirds of her calories) will come from fresher foods.

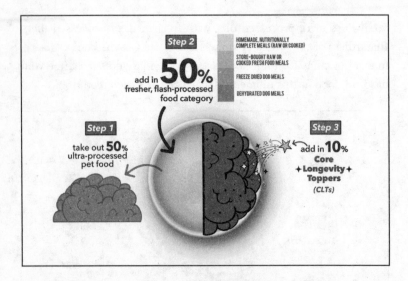

100 Percent Fresher Food Bowl Upgrade: If you choose this stellar option—the gold standard in Longevity Junkie circles—you've made the decision to eliminate *all* ultra-processed dog food from the bowl—bravo! Your dog will receive 100 percent of her calories from the healthiest pet food category: the fresher, flash-processed

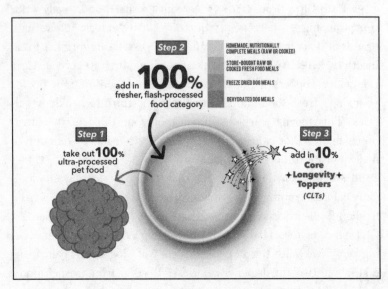

category. Your dog will also enjoy up to 10 percent CLTs to supercharge her bowl on a daily basis. As you know by now, the goal is to feed the freshest, most nutritionally dense food your budget and lifestyle allow; you can't do better than 100 percent!

Of course, these percentages are just suggestions. And remember, you don't have to choose just one fresher food category. Many people find hybrid, mix-and-match fresher food plans work best for their lifestyle, making homemade meals when they can, using freeze-dried foods when they're camping on the weekends, and providing commercial raw or cooked foods during the week. If you're already feeding fresher foods, you may simply need to work on improving variation between recipes, brands, and protein sources, which diversifies the microbiome and nutrient spectrum. If your dog isn't accustomed to eating a variety of different foods, introduce new foods and brands slowly, allowing plenty of time for her body and microbiome to adjust. Once she's accustomed to eating a variety of new foods, you can mix and match different types of foods, depending on your schedule, budget, and freezer space.

The first example of a Forever Dog Meal Plan is a 100 percent fresher food meal swap three times a week (noted by check marks), maybe a homemade meal on your shortest workdays. The remaining meals have up to 10 percent CLTs mixed in to provide superfood fuel.

Example 1

PET FOOD MEAL SCHEDULE 🐾

MON	TUE	WEN	THU	FRI	SAT	SUN
meal 1	meal 1	meal 1	meal 1	meal 1 ✓	meal 1	meal 1
meal 2	meal 2 ✓	meal 2	meal 2	meal 2	meal 2	meal 2 ✓

The second example is a 50 percent fresher food swap for six of fourteen weekly meals. Some of those six hybrid bowls may be 50 percent raw food and 50 percent kibble, or 50 percent freeze-dried and 50 percent gently cooked. As you can imagine, the possibilities for mixing and matching are endless. The meals are topped off with CLTs.

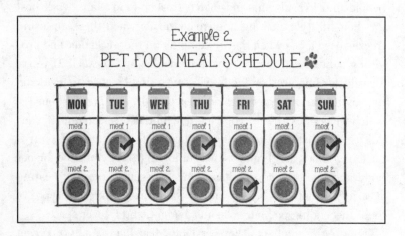

Improving your dog's bowl isn't an all-or-nothing approach. Start by upgrading a few meals a week. And start small. Is it okay to start by simply improving the quality of treats? You bet. Freeze-dried or dehydrated all-meat treats are far superior to ultra-processed, carby junk food snacks from the grocery store. If you own a dehydrator, you can dehydrate any of the fresh Longevity Foods for inexpensive, self-stable, DIY treats. Is it okay to take three months to work up to 25 percent fresher food in the bowl? Sure. Is it okay to make your first step a better-quality dry food? Of course. Just start somewhere. Wherever you feel comfortable.

Introducing New Foods

Add all new foods *very gradually* to your dog's diet. Keep your dog's base diet the same while you introduce CLTs, to allow some time for your dog's microbiome to adjust to brand-new foods in the form of 10 percent healthy "extras." This is especially important if his current diet is largely composed of ultra-processed food and/or he suffers from digestive disorders. Chances are his microbiome isn't diversified, and big food changes can cause big GI issues; *slow and steady wins the food diversity race.* We recommend introducing CLTs one at a time, as rewards (treats) or tiny food toppers, if your dog is sensitive. And be patient with your pup: if she declines a dime-size piece of jicama today, don't be dissuaded; try again tomorrow. This is a wellness marathon, not a sprint.

Pro Tip: A dollop of canned 100 percent pumpkin (or fresh, steamed pumpkin, if you can get it) on food helps firm up soft stools and eases the dietary transition for many dogs (about one teaspoon for every ten pounds). Alternatively, slippery elm powder from your local health food store works wonders for soft stools if you find you've transitioned too fast or your dog's snack causes loose bowels. We call this nature's Pepto-Bismol. If your dog has diarrhea, activated charcoal (also from your health food store) to the rescue! One capsule for every twenty-five pounds of body weight usually does the trick. Wait until stools are 100 percent normal before introducing more new foods.

EXTRA SUPPORT FOR SENSITIVE STOMACHS

Probiotics and digestive enzymes added to your dog's current food (before you diversify his actual food) help pave the way and prepare the GI tract for a more seamless transition to new foods and nutrients. These supplements decrease digestive stress in dogs prone to being gassy and having GI upset. Probiotics (as discussed in Chapter 8) are beneficial bacteria that keep the GI tract in balance while digestive enzymes assist in the digestion and assimilation of food. There are many brands of dog digestive enzymes that supply additional sources of amylase (to digest carbs), lipase (to digest fat), and protease (to digest protein) at your local independent pet retailer (or online). Rotate through different brands and products for the most supplement diversity over time.

Variety Is the Spice of Life

Diversifying your dog's diet fundamentally means diversifying her spectrum of nutrients and her microbiome, which does great things for a dog's overall immune system. Whether you're adding fresh herbs and spices, exploring a new protein source for treats, or trying out a brand-new type of dog food, your dog's taste buds and body are ready for a new adventure. How often you change up your dog's meals, proteins, and recipes depends on your dog and your lifestyle. Some people feed their dog different meals every day, much like they feed themselves. Some folks change proteins and brands every other bag/box, every month, seasonally, or every quarter. As far as switching things up, there is no right or wrong schedule. Look at your schedule and your dog's needs, and go from there.

If your dog is finicky or has a sensitive gut, you'll take more time introducing new foods and brands. If your dog has a specific health challenge, say food allergies or irritable bowel syndrome (IBS), you may find a handful of snacks, toppers, proteins, brands, and recipes that work exceptionally well managing your dog's issues; rotate among the things that you know work. Note in your Life Log those foods your dog likes and those she may initially snub—it's worth many tries or another presentation (try gently steamed in lieu of raw the first few times) when you first introduce new foods. Have fun experimenting with foods, food schedules, and Forever Dog Meal Plans; customize your approach around you and your dog, and please don't compare yourself or your dog with others—

you're both unique, and your feeding philosophy and approach are uniquely yours.

CREATE A CAROUSEL OF CANINE CHOICES

Your dog won't like all of the foods suggested here, but that's the fun of it—we guarantee you'll both enjoy the journey of discovering the foods she *does* prefer. Your food journey with your dog will be a delightful discovery mission, as you work together to discover her unique food preferences. When you offer tiny bites of new fresh food choices, you're stimulating your dog's senses and engaging his brain. Even if he decides the tidbit isn't his favorite, keep offering dog-safe foods from the fridge. You two are on a mutual journey of gustatory discovery that will last the rest of your dog's life!

Focus on Feces

Poop is a wonderful barometer of how well the GI tract is responding to new foods (and how healthy her gut is). We recommend monitoring your dog's poop on a daily basis to assess and calibrate the speed with which you implement new CLTs or a dietary transition. If stools soften, slow down the pace and decrease quantity of your offerings. Each dog is different, and it's important to understand and honor your dog's own physiology. If your dog has never had any fresh, new foods in his diet, he may not even show an interest in many of the CLTs. Don't dismay; instead try other fresh foods on the list until you find one or two he likes, then slowly diversify your offerings at a pace his brain and body can manage. As your dog expands his palate, you'll learn his preferences and even see them change over time.

If stools are stable and you're ready to upgrade your dog's bowl with a diet change, poop quality helps determine how quickly to increase new food and decrease old food. As a general guideline for healthy dogs, when transitioning to a brand-new diet, replace 10 percent of your dog's current diet with 10 percent new food. If poop is fine the next day, gradually increase the amount of new food by 5 to 10 percent each day, swapping out more and more old food with the new diet, until your old food is gone. If you encounter a soft poop, don't increase the volume of new food until the stool firms up, then continue your transition. If you've done Pet Food Homework and decide to replace your current brand with a healthier choice, buy or make the new food long before your old food runs out so there's no digestive upset from a too-hasty transition. It's never a good idea to simply run out of old food and then begin a brand-new diet; the body does best with a transition period that allows the gut microbiome to adjust.

Once your dog is eating his new diet and stools are good, you can begin the enjoyable process of finding the next brand, recipe, or new protein to introduce. Over time, as the microbiome diversifies and becomes more resilient, most people find they're able to bounce from protein to protein and between brands and types of fresher diets without any GI repercussions in their dogs, much like healthy-gutted humans can eat a wide variety of foods every day and not have GI consequences. Variety is the spice of life, not just for us but for the microbiome and nutritional benefits of the entire animal kingdom.

Most important, create a Forever Dog Meal Plan that works for your lifestyle. A lot of Longevity Junkies find great enjoyment and gratification in providing a variety of different recipes—both homemade meals and commercial fresher diets—several times a week. Other people don't have the emotional, temporal, or financial bandwidth to plan beyond simply buying a different brand (and protein) once they're halfway through their current stash of food. When they're halfway through their current bag, they head back to their local pet store and buy a different brand made with a different type of meat. They mix their current food with the new food, fifty-fifty, until they run out of their current dog food; when they're halfway through their new bag, they repeat the process. In essence, they're diversifying their dog's microbiome by changing brands and flavors with every bag, plus adding in CLTs and whatever is in their fridge that's appropriate. This is a perfectly acceptable way to diversify your dog's nutrient intake. Do what works best for you.

DRAMATIC ADDS TO DRY FOOD

Strawberries, blackberries protect against oxidative damage caused by mycotoxins.

Carrots, parsley, celery, broccoli, cauliflower, brussels sprouts reduce carcinogenic effects of mycotoxins.

Broccoli sprouts inhibit AGE-induced inflammation.

Turmeric and ginger mitigate damage caused by mycotoxins.

Garlic reduces tumor incidence from mycotoxins.

Green tea reduces DNA damage from mycotoxins.

Black tea protects the liver from mycotoxin damage and inhibits AGE formation in the body.

If your ex-husband has the dogs for the weekend and only feeds kibble, don't hesitate to feed fresher meals during the week, when the dogs are with you. We can't say it enough: Food can heal and food can harm, and you have tangible tools to make wiser choices, but don't allow this knowledge to stress you out. The goal is to nourish your dog to the best of your ability, armed with enough knowledge to provide healthful variety and to dissipate any underlying concerns. And just like us, dogs can eat some "fast food" and be fine; the key is not to make a habit of living on ultra-processed food as your primary source of nourishment.

Portion Control and Volume of Food to Feed

If you are not changing your dog's base diet right now and your dog is at his optimal body weight, you don't have to change the number of calories you're feeding; but you do need to keep tabs on your dog's eating window (ideally eight hours). If you choose to upgrade your dog's food and switch to 25 percent, 50 percent, or 100 percent (or any amount in between) of fresh, flash-processed food, you need to calculate how much new food your dog will be eating, based on the calories in the food (not the volume or portion of food).

How do you know how much new food to feed? You probably already know the *volume* of food you're feeding (like one cup twice a day), but you may not know how many *calories* your dog is currently consuming. You can find calorie information on your dog food bag. Because every food is different, it's impossible to simply swap one brand for another—they're all different calorically, and sometimes by *a lot.* You know how many calories your dog consumes daily right now, so you can calculate the volume of new food needed to maintain his current weight. In short, your dog needs the same number of calories he's been eating to maintain his weight, but foods don't have the same calorie count, so your math is important.

HOW TO CALCULATE CALORIES
WHEN SWITCHING DOG FOODS
───────

You can find calorie information on the food bag. As an example, if your dog's current food is 300 calories per cup and she eats two cups a day, she's consuming 600 calories a day. If you decide to feed 50 percent fresher food, then 50 percent of her calories will come from her new food, so 300 calories from old food + 300 calories from new food = 600 calories a day. If her new food contains 200 calories per cup she will eat 1.5 cups of new food a day (300 calories) + 1 cup of old food (300 calories). You can calculate your dog's daily baseline calorie requirements by multiplying her weight in kilograms (kg) by 30, then adding 70. For a 22.7 kg (50-pound) dog: 22.7 x 30 + 70 = 751 calories needed a day. This equation doesn't factor in calories needed for rigorous exercise, so scale up or down, depending on activity level.

Mixing and Matching Myths

The urban legends about mammals' (both dogs' and humans') inability to digest cooked and raw foods together in the same meal abound. We've heard so many crazy myths over the last decade that we are devoting an entire paragraph to dispelling these unfounded rumors. Quoting board-certified veterinary internist Dr. Lea Stogdale: "Dogs are physiologically adapted to eat *everything*: raw or cooked, meat, grain, and vegetables . . . sometimes after rolling in it." Research conclusively demonstrates there are no negative digestive consequences (for humans or dogs) when consuming both raw and cooked protein, fat, and carbs in one meal. A healthy human can eat a salad (raw veggies) with croutons (cooked carbs) and a chicken breast (cooked protein), or a sushi roll (raw protein and cooked carbs) and seaweed salad (raw veggies) without digestive confusion (aka vomiting or diarrhea). Likewise, healthy dogs can consume raw and cooked foods in the same meal (and have been for thousands of years); digestive studies confirm they effectively absorb fats, proteins, and carbs (sugar) just fine when mixed together in one meal, just like us. If you partition or stagger the consumption of your own food (eating cooked and raw carbs, fats, and proteins in a certain order) and feel compelled to do this for your dog, it's okay but unnecessary; dogs eat poop and lick their backsides and have much more resilient GI tracts than us. If your dog has a history of pancreatitis or a "sensitive stomach," add digestive enzymes and probiotics to help your dog process the newly introduced foods.

Respect the Power of When

Remember: Science says *when* you eat is as important as *what* you eat. **These are the two most important factors that dictate life span and health span.** If changing what you feed your dog today feels overwhelming or isn't feasible, just start with the timing. In our conversations with Drs. Satchidananda Panda and David Sinclair, they heartily endorse the routine caloric restriction (feeding a set number of calories each day) and timing meals to match the body's inherent circadian rhythm and maximize metabolism. "Every hormone, every digestive juice, every brain chemical, every gene (even in our genome) goes up and down at different times of the day," Dr. Panda reminds us. He also highlights that the gut microbiome follows the circadian rhythm of the body; when we go without food for several hours, for instance, that produces a different environment in the gut. So a different set of bacteria will flourish and are useful for cleaning up the gut. Adopting consistent and strong eating and fasting rhythms fosters a different set of gut bacteria; the composition of our microbiome goes through a daily change, for better or worse, depending on what and when we're eating in relation to other cues like sunlight and ebb and flow of hormones.

Dr. Panda is a big proponent of time-restricted feeding, and he uses a common analogy to drive home the message: Except for nocturnal animals, animals don't eat in the dark—dogs don't hunt at night, which means when the sun goes down it's time to stop eating. Problem is, some dogs live in homes where the blinds are drawn all day, and they can't readily discern when the sun comes up or down. Panda prefers time-restricted feeding (creating an eating window) over intermittent fasting because the fasting route can encourage cheating—consider, for example, fasting in the morning and waiting to have lunch as your first meal, but then gorging at night before bed. Eating before bedtime is not ideal for the circadian rhythm.

When you understand and then honor your dog's circadian rhythm, the list of benefits is profound: enhanced resilience, improved reproductive health, improved digestion, better heart health, improved hormone balance, reduced depression, improved energy levels, reduced risks for cancer, reduced inflammation, reduced body fat, reduced hypertension, improved motor coordination, reduced gut distress, improved blood glucose, improved muscle function, increased longevity, reduced infection severity, improved brain health, improved sleep, reduced risk for dementia, reduced anxiety, and improved alertness. The list goes on, but you get the point!

Dr. Mattson's lab at the National Institute on Aging has confirmed these findings: Mice who eat within an eight-to-twelve-hour window live longer than mice with unlimited access to food, *despite consuming the same number of calories.* A rudimentary way to remember the facts: When our circadian clock says the body is prepared to eat, the food is healthy; when the clock says no, that same food can actually do harm to the body. Respecting the circadian rhythm of our dogs will go a long way in keeping them healthy and preventing illness in their old age.

Dr. Sinclair echoes these recommendations. When we point-blank asked him, "From everything you've learned, what lessons do you apply to your own dogs?" his simple but powerful answer was: Be as lean as possible, don't overfeed, and exercise a lot. "It's okay to be hungry," he states frequently. Think about it: Neither ancient humans nor our dogs' ancestors had the luxury of multiple meals and snacks a day, and they certainly didn't break the fast every morning with a bounty of food on an exact schedule. Dogs consumed the prey they caught and then fasted until their next success.

Our modern meal practices are more a product of culture and habit in the land of plenty than anything else. **When you honor your dog's natural circadian rhythm, you optimize his health.** Simple as that.

You can choose among a variety of time-restricted feeding (TRF) strategies. Start by creating an "eating window." If you are one of the rare folks who leaves a bowl of food down for your dog all the time, your first step is to pick the bowl up. The days of the all-you-can-eat buffet are done. We think that's actually the definition of heaven, so while we are here on Earth, we have to abide by earthly rules and physiological principles, which include honoring your dog's physiology. Fido is a canine, not a goat! Ruminants and other vegan animals (cows and horses, etc.) must snack all day; they are massive animals that eat grass for energy and sustenance, and they need to eat *a lot* of grass to maintain 1,000 pounds of body weight. Their physiology, from their wide, flat molars (for chewing, chewing, chewing) and extra-long GI tracts needed to ferment the energy out of all that grass, requires them to nibble almost constantly to fuel their massive metabolic machinery. Dogs are the exact opposite.

Vets regularly recommend fasting for certain dog illnesses, including reducing toxic side effects and improving the benefits of chemotherapy and acute vomiting and diarrhea. But only wellness vets understand the benefits of TRF for healthy dogs and institute it in practice. We'll leave actual fasting protocols to you and your vet, but time-restricted feeding isn't fasting. It's feeding the *same calories your dog would normally eat, but during certain hours of the day*.

Practice what we call targeted calorie consumption within a certain window of time, ideally *eight hours* for most dogs of normal weight, and cut off all calories at least two hours before bed. Targeted calorie consumption within an eight-hour window is our way of saying time-restricted feeding, but it sounds gentler and you're not really "restricting." You're simply being strategic and calorie conscious.

We recommended TRF to hundreds of Longevity Junkies, and the feedback has been incredible: Yes, everyone in the house is sleeping better; dogs are less anxious during the day; digestion is better and sleep is more sound at night. But most important, TRF also produces all sorts of health benefits, which have been noted since implementing TRF and without even changing their food. **You can positively affect your dog's metabolism and overall well-being simply by feeding all her daily calories in a defined period of time!**

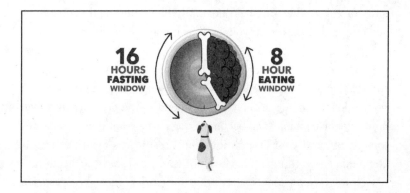

We get it, making the conscious choice to skip treats after dinner, especially if your dog is used to snacking well past dinner, can be a challenge. This is a fantastic time to swap those after-dinner snacks for an after-dinner walk. If your dog is used to a specific food-related ritual after dinner, replace his usual snacks with bone broth ice cubes (recipe on page 242). And if you get home too late for dinner, simply let Fido skip a meal. In fact, if your healthy dog ever signals he doesn't want to eat a meal, let him have his way. **There's nothing wrong with skipping a meal; it's a therapeutic mini-fast.** Rodney's dog, Shubie, regularly decides on her own to fast for more than twenty-four hours, letting him know thirty-six or even forty-eight hours later when she's hungry and ready for her next meal. If your dog naturally doesn't want to eat breakfast, for instance, allow him to fast until he tells you he's hungry. Feed the first meal when your dog tells you she's hungry: the beginning of her eating window.

If your healthy dog doesn't care or pay much attention to the number of meals she eats a day, then feed her once a day, when it's most convenient for you (but *preferably at least two hours before bed*). If you feed three meals a day, divide the middle meal portion between the first and last meal and begin feeding two meals a day. If your dog is regimented to eating "lunch," play a quick round of fetch or tug at the time he'd normally be eating; he'll probably be so delighted to play with you, he'll forget about lunch. You can also use Longevity Foods and CLTs as rewards in treat-release puzzles or a snuffle mat during the time your dog typically expects to be fed. Your dog may beg or otherwise share her opinions with you about this new regimen. Do not bend to the pressure, however adorable (or irritable) she becomes: dogs are evolutionarily adapted for fasting, and your "tough love" will result in a healthier pooch. Remember two things: They are not cows, and they *will* acclimate. Two low-glycemic meals a day allow for a brief (*but highly beneficial*) period of digestive rest in between periods of digestion.

Within the eating window, most of the experts we interviewed also suggested mixing up the *time* your dog eats on a regular basis. Changing up mealtimes enhances metabolic flexibility. If you're typically rigid about mealtimes, start by feeding your dog half an hour earlier one meal, then fifteen minutes later the next. This strategy works well for dogs who produce stomach acid on a schedule, like clockwork, and vomit bile if they're not fed at an exact time. By gradually changing up mealtimes, using CLTs as training treats throughout their eating window, and ignoring the begging, you are conditioning your dog to be more metabolically flexible and activating all the longevity benefits of time-restricted eating, all without changing your dog's caloric intake.

The Body Score Test Sets Your Dog's Eating Window

Underweight dogs: Determine (with your vet, if that makes you more comfortable) your dog's *ideal* body weight and the calories necessary to maintain that weight. Feed all those calories in three separate meals within a ten-hour eating window. Once optimal body weight is achieved, divide calories into one to two meals per day to maintain her weight.

Lean, average animals (ideal body weight): Feed all calories in one or two meals, within an eight-hour window to maintain optimal body weight.

Healthy (non-diabetic) overweight/obese: If your dog needs to lose a lot of weight, ask your vet to help you set progressive and safe weight loss goals. Aim for a 1 percent weight loss each week. For example, a fifty-pound dog that needs to lose ten pounds should lose about half a pound per week, or two pounds per month. Weigh your dog weekly to confirm you're on track. For the first two weeks, feed all calories in a ten-hour window (this is the optimal eating window that has yielded success for humans with metabolic syndrome and has been successfully duplicated in animal models). Then reduce your dog's feeding window to eight hours. You can divide her calories into however many meals you wish (most pet parents choose three, so that their dogs eat smaller portions more frequently). Once

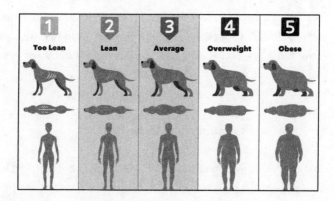

your dog achieves his ideal body weight, continue feeding all calories in an eight-hour window, divided into one to two meals.

If you have an obese dog, when you calculate the cost of switching to fresher food diets, remember to calculate the amount of food needed to maintain your dog's *ideal* body weight, not the inflated price you'd spend to maintain your dog's obesity. We've seen many people who could dramatically improve the quality of their dog's diet without actually spending significantly more by making the decision to feed a far smaller volume of superior-quality food. Smart choice.

THE COMPELLING CASE FOR ONCE-A-DAY FEEDING

The scientists and researchers we interviewed agreed one meal a day for healthy dogs is ideal, in terms of maximizing autophagy and minimizing metabolic stress. Dr. Panda stressed that consuming all calories in an eight-hour feeding window, whether one big meal or six smaller meals, is the most important tool to maximize longevity benefits. Dr. Fung stressed that each time our dogs eat a meal, their bodies emerge from rejuvenation mode and shift into digestion mode, so **fewer meals enhance the amount of rejuvenation and autophagy occurring**. We feed our dogs once a day, to maximize the health benefits that materialize only in rejuvenation mode (when they aren't eating). Dr. Panda offers the most important advice: **Limiting our dogs' eating window to less than ten hours a day (and ideally eight) is the most important strategy, regardless of how many meals you choose to feed**.

Treat in a Timely Manner

Think of treats like human snacks: What you choose to snack on, how much, and how often all factor into your overall health and well-being—a lot. If you never snack and your dog never snacks, feel free to skip over this information (but we suspect this section applies to the majority of us).

Ideally, we should be using treats for a purpose, as a reward—to strategically communicate with your pup, such as "job well done!" If you give treats because she's cute and you love her, we understand (they *are* cute and we *do* love them), but we encourage you to reduce the size and frequency of those treats, replacing those calories with cuddles, kisses, play, and walks. Treats fed as snacks (or "just because") can turn off autophagy, if fed too often, in too big portions, or at the wrong time. One of our suggestions for treats is using tiny bits of your dog's food allotment as rewards: boring, we know, but functional.

Becoming our dog's emotionally rich food substitute: Sometimes we use food as a replacement for being emotionally present. Presence—mindfulness—is critical to human health and well-being. As you begin replacing junky dog treats with smaller, less frequent healthy morsels, also replace those empty calories with purposeful and thoughtful engagement with your dog. Put your phone away. Look at your dog. Talk with your dog. Be in the moment. Take a few minutes to schmoodge on your dog—you'll *both* get a boatload of oxytocin. Mindfulness is a powerful elixir for both you *and* your dog.

As dog trainers and behaviorists know, there are lots of important and viable reasons to reward your dog with small bits of food, most notably for training purposes and to reinforce behaviors we want (like going potty outside or mastering a new trick). Our treat/reward goals are to feed tiny morsels, ideally the size of peas (or smaller), to avoid creating an insulin spike. For the healthiest treats, chop up any of the Longevity Foods listed in Chapter 7 and use them as rewards. You can also use your 10 percent of CLTs as training treats (versus putting the goodies directly in the bowl). Blueberries are perfect-size treats for big dogs. One thinly sliced organic mini-carrot yields four to six training rewards. Two mini carrots, sliced into thin rings, can supply an entire day's worth of rewards when you're training your dog and need to communicate "great job!" None of the Longevity Foods listed disrupted our dogs' blood glucose when we used them as training rewards throughout the day; we verified this with a glucometer.

Think about everything besides dog food that goes into your dog's mouth. Do you habitually give your dog the corner of your sandwich or the last bite of pizza crust? Stop. It's fine to share some human food with your dog, but make it biologically appropriate, which means nix the carbs. Share bites of clean meat, fresh produce, seeds, and nuts. Ditch the ultra-processed treats in the cupboard. Thankfully the pet industry is brimming with biologically appropriate freeze-dried and dehydrated all-meat dog treats. If you're going to use store-bought treats in addition to Longevity Foods, make sure to read the label. Single-ingredient, all-meat or veggie treats are what we recommend because they're low-glycemic, with no added fillers or preservatives, and you can find human-grade, organic, and free-range options. Treat labels should be simple and understandable, such as "dehydrated free-range rabbit," "freeze-dried lamb lung," or "beef liver, blueberries, and turmeric." Fresh, flash-produced treats with minimal ingredients are a smart choice. You should be able to break treats into pea-size pieces. Remember, anything your dog eats in addition to his meal(s) are "extras," and those freebie calories (10 percent) should count for good! Using tiny amounts of healthy, fresh food as treats will not disrupt circadian rhythm or create metabolic stress.

Once you find some fresh food treats your dog loves, stick with them, but continue experimenting with new foods—learning about your dog's expanding palate will be fun and exciting for both of you. If your dog has never been offered any fresh food and appears uninterested or confused, don't panic. Many dogs that have never had the opportunity to eat fresh food don't understand what it is, initially. As their taste buds respond to new and different foods, they'll be more likely to try new foods they may have refused previously, so don't give up.

Don't know where to begin? If you're on a tight budget or tend to be cautious, continue feeding the same food you've always served for now, but focus on better meal timing. Create an eating window and make sure your dog is consuming the ideal number of calories to sustain a healthy lean weight. Fix your treat situation. Add in CLTs. Begin or diversify your daily exercise routine. Plan daily sniffaries and other environmental enrichment opportunities (aka "doggie happy hour"). Optimize your home environment. Clean up your air. Ditch chemical cleaners. And decrease stress by enjoying exercise and an enriched social life.

If you're ready to step it up but not quite prepared to go all out, you can improve your dog's diet incrementally by blending better-quality: Go from "good" to "better" brands, less-processed food with fresh food (if you were at 25 percent fresher foods, bump up to 50 percent). Give attention to better food timing, and adding in as many Longevity Foods as possible.

For Longevity Junkies ready to take the plunge into Forever Dog territory, discontinue processed food, feed all real food, focus on better food timing, and add in as many Longevity Foods as possible while simultaneously optimizing your home environment and decreasing stress by getting exercise and enjoying an enriched social life.

LONGEVITY JUNKIE TAKEAWAYS

➤ Getting Started (Step #1): Introduce Core Longevity Toppers (CLTs)
  ➤ In addition to CLTs, decide which treats you will try. Here's a quick summary of some of our cube-able or naturally bite-size favorites: blueberries, peas, carrots, parsnips, cherry tomatoes, celery, zucchini, brussels sprouts, apples, sunchokes, asparagus, broccoli, cucumbers, mushrooms, green bananas, berries, coconut, tiny bits of organ meats, and raw sunflower and pumpkin seeds.

➤ Getting Going (Step #2): Evaluate Your Base Diet and Freshen Up
  ➤ Complete your good/better/best Pet Food Homework: count carbs, do adulteration math, calculate synthetic nutrition addition.
  ➤ Choose your category of fresher food(s): homemade raw or cooked; commercial raw, cooked, freeze-dried, or dehydrated; or a combination.
  ➤ Set your fresh percent: 20 percent, 50 percent, or 100 percent fresher foods.

➤ Keep notes in your Life Log about successes, failures, and new ideas.

➤ Introduce fresh foods and CLTs one bite at a time and transition your dog onto his new diet slowly, at a pace that doesn't create diarrhea.

➤ Do the body score test and make sure you're feeding the right number of calories to achieve or maintain your dog's ideal body weight.

➤ Choose your dog's eating window (with an eight-hour goal) and number of meals—noting there's nothing wrong with skipping a meal if your dog is healthy. Stop eating at least two hours before bed.

10

Putting the OGS into DOGS

Fitness Guidelines and Controlling for Genetic and Environmental Impacts

> Whoever said diamonds are a girl's
> best friend never owned a dog.
>
> —Unknown

Darcy, a twenty-one-year-old tiny mixed-breed dog (we had the pleasure of wishing him a happy birthday via a video chat), ate a balanced meal once a day. Exactly what Dr. Panda says to do. Darcy's parents made many other wise lifestyle choices that also contributed to his success in life. His parents attribute his long life to eating home-cooked meals since he was seven years old; for two-thirds of his life he ate a human-grade, low-carb, freshly cooked diet, with added fresh salmon and splashes of perna mussel, turmeric, and apple cider vinegar. Some days he chose to fast, and his parents let him fast for as long as he wanted to, sometimes skipping more than one meal.

When Darcy was younger, he spent the majority of his days outside in the yard with his spaniel mix brother. They had access to lots of healthy dirt, fresh air, chemical-free grasses to graze on, and various types of environmental stimulation and enrichment. His parents told us he was not exposed to routine veterinary or home chemicals. He was immunized as a puppy but did not get annual vaccines later on in life (you'll learn about using vaccine antibody titers to determine if your dog actually needs more vaccines after adulthood later on). Later in life when he started getting stiff and slowing down, he had hydrotherapy (water exercise) to help keep his joints and muscles moving well without high-impact exercises. Darcy's guardians followed the Forever Dog principles and he lived a long and happy life.

Now that we've stuffed you with dietary rules and suggestions, the time has come to finish our Forever DOGS Formula with the final three aspects to building a most durable dog:

- ➤ **O**ptimal movement
- ➤ **G**enetic considerations
- ➤ **S**tress and environmental impacts to manage

Let's get to it.

Optimal Movement

As we write this book, Germany is drafting a law specifying that dogs must spend time outside of confinement (be walked) two hours a day. One thing all the Forever Dogs we met had in common was *a lot* of *daily* physical activity. All dogs are natural-born athletes (minus the breeds that can no longer breathe or move normally). Most rehab vets and physiotherapists believe dogs do best when they run or sprint—off leash—at least once a day, *in addition* to aerobic conditioning (exercise). Even better, swimming allows dogs to move their bodies fluidly and in a way that is natural for them, putting all of their joints through a wider range of motion, which can't happen when they are on a leash.

Dr. Enikő Kubinyi, the head researcher of the Methuselah dog study, told us that twenty-seven-year-old intact female Buksi and twenty-two-year-old Kedves lived a "free life": They could make choices based on preferences, they weren't constantly restrained in their movement, and both dogs spent lots of time outside. She noted the oldest dogs from Australia, Bluey and Maggie, lived similar lifestyles, all spending ample time outside on a daily basis. Other interesting similarities: all of these dogs consumed some raw, unprocessed foods; sampled grasses and plants from their environment; and had modified vaccine and veterinary pesticide schedules.

City dogs might think they get to live the high life with their urban owners, but studies show they are more likely to be more sedentary, endure more stress with increased cortisol levels and behavior issues, have poorer social skills, and enjoy less contact with dirt and immune-boosting microbes, among other effects. Let's face it: If you live in a city or the suburbs, you most likely have a more stressful life defined by long work hours spent under artificial light. Their pets are unable to choose what they sniff and how long they get to move their bodies because they're (hopefully) taken out on well-confined walks on pavement for limited amounts of time.

City Life Means Creative Exercise Protocols

As Ingrid Fetell Lee writes: for eighty thousand generations, nature is not a place we went, it was a place we lived (with animals). Only six hundred generations have passed since the agricultural revolution gave rise to permanent communities and only twelve generations have passed since the birth of the modern city, with all of its concrete and profoundly deficient green spaces. Dogs have lived in cities less than six generations, so we need to cut them a little slack. Whether you train your dog to run on the treadmill, drop her off at doggie day care, hire a dog walker, run the stairs of your apartment building, sign up for underwater treadmill cardio, or find an empty baseball field to fetch the frisbee after hours, there are many creative ways to meet your dog's daily physical exercise requirements, even in a concrete jungle. Don't let your lack of imagination deprive your dog of the daily movement she needs to be physically and mentally balanced!

The reality is most dogs don't get enough exercise and aren't given the opportunity to move as much as they'd like. This leads to pent-up energy, which can lead to hyperactivity, increased anxiety, and destructive behaviors—some of the top reasons dogs end up at shelters. There's solid science behind the notion that a tired dog is a good dog (just as parents know that physically tired kids are good kids). Sometimes our clients ask how much they should exercise their dog every day, to which our simple response is "as long as it takes to make them exhausted by bedtime." Although there are some basic guidelines to exercise, like there are for humans, generally speaking, dogs need *a lot* of daily aerobic exercise to be mentally and physically fit, much more than us—and that's part of the problem.

We've interviewed the guardians of some of the oldest dogs in the world. Augie's dad reported she swam an hour a day, even at fifteen years of age, and reached her second decade; toward her death in the spring of 2021 she mostly walked for exercise. According to Brian McLaren, thirty-year-old Maggie followed him on his tractor and ran three miles, from one end of the farm to the other, twice a day, seven days a week, for two decades. That means she banked an average of *twelve and a half miles of exercise a day.* The one consistent variable among all of the oldest dogs in the world: daily rigorous exercise, rain, snow, or shine. Ann Heritage writes that Bramble, her twenty-five-year-old dog, walked several hours daily. In Mongolia, where Bankhar dogs roam with humans who live a nomadic life, they are known to work strenuous jobs on the land to protect livestock even as old as eighteen years. These unique dogs are large, athletic, protective, and need comparatively little food for their size (another clue to the benefits of eating lean).

We all know exercise is good for us. We won't even go into the human studies on the value of exercise, other than to say there's even research to show that pet owners feel better and happier when they walk their dog. The body of research and evidence about how exercise can dramatically improve a dog's health and life (and attitude and behavior) is overwhelming. We covered a lot about the benefits of an active lifestyle in Part I, but here's a list of science-backed conclusions about dogs:

> Reduces fearfulness and anxiety
> Decreases reactivity and increases good behavior (i.e., reduces or eliminates common boredom-induced behavior problems)
> Ups their threshold for noise pollution and separation anxiety (they can tolerate more)
> Provides a means of lymphatic detoxification (the lymph system is an important part of the immune function, so keeping it clean and healthy is key)
> Lowers risk for all manner of ills, from overweight and obesity (and helps bring those conditions under control) to joint disease, heart disease, and neurodegenerative disorders
> Maintains a strong musculoskeletal system, critical for carrying dogs into their geriatric years
> Helps normalize and regulate the digestive system
> Boosts production of that star antioxidant glutathione and also increases meaningful AMPK, the antiaging molecule
> Helps manage blood sugar and reduces risk of insulin resistance and diabetes (Tip: Even a quick ten-minute walk after each meal can decrease blood sugar spikes)
> Builds confidence and trust while improving a dog's ability to be calm

The more hyperactive and excitable your dog is, the more she needs to move. Dogs with anxiety and stress are able to bring their stress hormones back into more healthful parameters with rigorous cardiovascular activity. All dogs need to exercise, regardless of athleticism, size, age, or breed. But most dogs don't get enough exercise, which is why there are so many overweight, achy-jointed, bored-to-distraction dogs today. And many older dogs get the short shrift;

geriatric dogs need more time to smell because their body and their other senses are not what they used to be. Giving older dogs ample time to smell outside every day is critical not only for exercise but for their enrichment and engagement with the world.

Most of us could use more physical activity every day, too. **Dogs should get a bare-bones minimum of twenty minutes of sustained heart-thumping exercise a minimum of three times a week to prevent atrophy; most dogs can benefit from longer, more frequent sessions. There is nothing wrong—in fact, there is everything right—with thirty minutes or an hour of exercise six or seven days a week. More is better!** Dogs throughout history have hunted, played, mated, and raised their pups outdoors, running miles and miles each week. Their days were active, social, and enriching, keeping them mentally and physically occupied all the time. Dog spend 60 percent less time resting when they are surrounded by other canines. Both humans and dogs are social creatures, motivated by others to be active. But dogs, like humans, need a little nudge sometimes. As the pet parent, you have to make sure your canine receives companionship, encouragement, and motivation to get outside and play, and that means scheduling long walks or play sessions. Without habitual opportunities to participate in aerobic exercise, even if they are within a normal weight, they could develop arthritis and other conditions affecting their joints, bones, muscles, and internal organs. Without habitual physical and mental challenges, your dog's behavior and cognition will deteriorate. This leads to "bad" behaviors like inappropriate chewing, jumping on people, scratching, digging, predatory play, nosing around for trash, disruptive barking and play, heightened reactivity, aggressiveness, and attention-seeking behaviors.

Dogs thrive when given the opportunity for daily "movement therapy," incorporating a wide variety of activities and exercises that puts all of his joints through their natural range of motion, building muscle tone and strengthening tendons and ligaments.

Consistent daily exercise has profound long-term health benefits that are a prerequisite for a maximal health span. One of the biggest issues we see as dogs age is a loss of muscle tone, which sets the stage for weakness, progressive degenerative joint disease, and decreased range of motion (not to mention injuries and increased pain, an underdiagnosed reason for aggression and behavior changes).

News flash: Being a pet parent is a seven-day-a-week job. Some people hope that they can make up for a work week full of inactivity with a weekend of fun and games. This may actually hurt your dog because sudden bursts of exercise can trigger an injury in a body that is not regularly conditioned to withstand movement. This could then lead to long-term joint damage.

Chances are your dog lies around most of the day waiting for you to get home from work. His tendons, muscles, and ligaments have also been lying around. If you come home from work and throw the ball twenty times, a likely consequence can be a torn cruciate ligament (the most common knee injury we see in veterinary medicine). And a few minutes of crazy play does not confer the same health benefits as thirty minutes of muscle-building, controlled cardio. Dogs can turn "on" in a second—they're just waiting for us to engage—and oftentimes don't have an "off" switch. It's up to us to warm up our dogs' bodies prior to rigorous play and know when to stop (by reading their body language). Most important, dogs do best when provided with daily opportunities to move their bodies and condition their musculoskeletal system in a way that's enjoyable to them. *All* dogs are wired for outdoor movement (even the tiny ones); they are designed to move their bodies, *a lot.*

The only way to prevent musculoskeletal atrophy with age is to move your dog daily; muscle tone doesn't come in pill form, and dogs need more of it as they age. This is especially important during your dog's midlife years, when you can really focus on building endurance and excellent muscle mass and tone to carry them healthfully into their geriatric years. Focusing on building a resilient musculoskeletal system in mid-life dogs is nothing short of accruing "frame insurance" for years down the road; this strategy serves large breed dogs exceptionally well. The goal is to create physically resilient bodies.

Since a big challenge many pet parents face is trying to squeeze in quality time with their canine BFFs, a daily exercise routine that includes your dog may be the ultimate solution. We may not feel up to it all the time, but most dogs are always ready to get up and go. Just know that a slow stroll with your pet is not enough of a workout. If you're going to walk, make it a power walk (with a pace of four to four and a half miles an hour, or a fifteen minute mile). Moving at this pace ensures that your pet receives cardiovascular and calorie-burning benefits.

Power walks are good for your pet and for you. They decrease your risk for obesity, diabetes, heart disease, and joint disease. However, before you can increase the speed of your walk, you may have to retrain your dog so he's not expecting his usual casual amble. We *love* those type of walks (aka sniffaris) for mental gymnastics, but they don't count as cardio. One day may not be enough to reprogram your pet. Moving from a sniff walk to a cardio session could take several weeks as your pet's endurance builds. Using different harnesses and collars is a great way to let dogs know what type of activity to expect; we use body harnesses and short leads for rigorous cardio and long lines with flat collars for leisurely sniffaris.

If walking at a fast clip is outside your ability, try other cardio exercises like swimming. The main reason I (Dr. Becker) opened an animal rehab/physical therapy center years ago was to give dogs a safe place to exercise in the winter. Underwater treadmills provide an awesome workout and are fantastic for older or out-of-shape dogs and dogs with disabilities. With a tutorial, tiny dogs can swim at home, in the bathtub. Many doggie day cares now offer treadmill services for larger dogs, and animal rehab professionals across the globe are trained and ready to help you develop a customized exercise protocol that meets your dog's specific needs. (Directories of rehabilitation professionals can be found on page 407 of the appendix.) There also are dozens of fun dog "sports" you both can enjoy; dogplay.com features organized exercise and play ideas for your dog. **Variety is important, but so is enjoyment. Choose activities from your dog's perspective,** making sure your choices resonate with your dog's personality and ability. As dogs age, their exercise regimens will change. Older dogs benefit from intentional muscle-strengthening exercises, like sit-to-stand training sessions (many rehab professionals offer teleconferencing sessions to teach you which exercises are best for your dog's needs). Regular at-home massage and gentle stretching sessions are a terrific way to help your dog feel great and for you to complete regular body scans, checking for lumps, bumps, and any other changes to your dog's body. The www.foreverdog.com site has more info about at-home exams and what to look for.

Brain games: Dogs need to engage their thinking skills *and* their bodies. Physical exercise is critical, but mental exercises keep dogs sharp (and prevents boredom) well into their senior years. Dogs who enjoy a lifetime of nose work (scent work), agility (or other sports), or brain-engaging puzzles have less cognitive decline as they age. Nina Ottosson and My Intelligent Pets make great dog brainwork puzzles, or you can create your own; we share some ideas on the Forever Dog website.

Just be sure to tailor your dog's workouts or play sessions to their body type and ability (for example, brachycephalic breeds have special breathing considerations), temperament (such as aggressive dogs), and age (older or with a physical limitations). The type, duration, and intensity of exercise you choose for your dog will be adjusted over time, but your dog should never stop moving.

Some breeds are predisposed to neurodegenerative diseases, and some dogs already have musculoskeletal trauma from accidents or injuries. It is especially important to create a customized exercise protocol for dogs with physical impairments that meets their individual needs, sometimes with the help of specialized harnesses and assistive devices.

CIRCADIAN-SETTING SNIFFARIS

Always open your blinds and shades before leaving your house in the morning. Don't leave your dog in the dark! According to Dr. Panda, animals living inside with dim lighting and closed blinds or shades during the day can become depressed, not to mention completely confused about whether it's day or night. He suggests a ten-minute walk in the morning and again at dusk, to allow our dog's body to produce the appropriate neurochemicals to wake up or wind down. The wise recommendation aligns perfectly with Dr. Horowitz's suggestion of allowing dogs to sniff to their hearts' content, at least once a day . . . to have a sniffari. As noted, we recommend **a circadian-setting sniffari *twice* a day, morning and night**. During a sniffari, your dog gets to choose what he wants to sniff and for how long: this is a *mental* exercise for your pooch, so no pulling on the leash! Allowing your dog to have generous opportunities to sniff is incredibly important for his mental and emotional health. (Plus research shows humans who take a fifteen-minute leisurely walk after meals curb risky blood sugar spikes all day!)

In addition to building more motion and momentum into your life with your pooch, we hope that you'll make a commitment to be your dog's wellness ally. You'll do this by following the D.O.G.S. principles in this book.

But first, take the pledge.

Take the Pledge

Stealthy physical and emotional surveillance is our responsibility, as guardians. You are your animal's advocate. Below is a pledge my (Dr. Becker's) friend Beth and I created many moons ago to remind pet parents of their awesome—and incredibly rewarding—responsibility. We encourage you to become your dog's wellness ally.

I am responsible for my health and well-being and for that of the dogs in my care. I will become a knowledgeable advocate for myself and my dogs in all realms of life. I understand that life, healing, and health are always changing, requiring me to learn and evolve in order for me to become an effective advocate. I will not abdicate this responsibility to any person or doctor. The emotional and physical health of my dog rests in my hands.

Keep your own at-home wellness records in your Forever Dog Life Log. Record your dog's weight every few months and the size and location of any lumps, bumps, and warts you find during your at-home exams. Ask for copies of physical exam findings, blood work, and laboratory tests so you can track changes in organ function. Log when any new symptoms start; note behavioral changes and what food and supplements each pet is taking. This ongoing health journal becomes invaluable when you're trying to remember when you changed foods, what day you gave heartworm prevention, or what month the increased water intake started. Leave this dog diary in a convenient place so you can quickly jot down notes. We use the Day One journal app on our phones for this purpose, because you can snap pictures and add voice memos easily.

A good question: How do you know the health of your dog on the inside? Laboratory tests are a great indicator. There are blood work panels for young and older dogs, as well as specialized diagnostics for Longevity Junkies who yearn for more information. Your dog may look happy and healthy and have few concerning symptoms, but that doesn't mean they don't need blood or diagnostic tests. Indeed, the metabolic and organ issues that plague your pet begin with "silent" biochemical changes, and tests can predict a diagnosis months to years before any symptoms begin. We can't tell you the number of times we've heard "I wish I had known sooner" when the diagnosis of kidney, liver, or heart disease is made. We *can* know sooner, thanks to modern technology, and you absolutely should leverage simple, noninvasive diagnostics to identify biochemical abnormalities *before* symptoms manifest—when we can actually do something.

Putting off testing until your pet is symptomatic may mean you're past the time when you can cure the disease or bring your dog back to good health. Proactive owners and veterinarians focus on identifying early changes on routine diagnostics that highlight the beginning of cellular dysfunction, before disease sets in.

Annual blood work gives you peace of mind that your dog's organs are functioning optimally. At some point in the aging process, normal values inevitably shift. Abnormal test results should be addressed and rechecked by your veterinarian. This is typically when owners seek out adjunctive support and/or second opinions. It's okay and oftentimes wise to add cooks in the dog health kitchen or add a concierge wellness service to help you navigate your animal's health concern to resolution. We would never expect one family practitioner (human or animal) to be capable of adequately attending to the progressive needs of aging family members. Seeking out a variety of veterinary perspectives and services for your aging dog is no different. (See our annual blood-work recommendations on page 399 of the appendix and newly released diagnostics at www.foreverdog.com.)

Genetics and Environmental Stress

We've provided a lot of information about the impact of lifestyle on health span, but we'd be remiss to neglect the power of genetics.

Genetics are especially consequential in dog breeding. The best we can do today to protect and promote healthy genomes is to reconsider how we breed our dogs; evolving our breeding practices is the only way to *ensure* a healthy genomic composition. We mentioned earlier that DNA screening is becoming increasingly more common, accessible, and useful in learning about our genetic predispositions and potential risks for certain ailments. DNA tests for dogs are also on the rise and will become more comprehensive in coming years. But the fact remains that a lot of dog breeding supports people's desires in how dogs should look at the expense of their health.

Although there are lots of great dog breeders who prioritize health over vanity, the demand for puppies fuels an industry easily corrupted by unscrupulous characters who are illiterate (and wholly uninvested) in breed wellness and are on stand-by to fill those misguided consumer needs. The swap of balanced brains and bodies for beauty has been devastating for many dogs. Preservation breeders have lost the battle to backyard breeders and puppy mills that have churned out tens of thousands of puppies in previous decades to fill a hungry pet-loving market; thoughtfully selected genetics and temperaments lost out to mass-produced, unhealthy litters.

The pandemic also has intensified exploitative breeding due to rising demands. We've seen a crush of online puppy scams, as people in perpetual isolation ache for the unrivaled companionship of a dog. Instead of investigating responsible breeders, however, many people just go to the pet store. Most (if not all) pet stores acquire their puppy supply from sources that don't prioritize genetic health; the same is true for the thousands of websites offering adorable, expensive, poorly bred puppies, delivered right to your door. The only way to avoid being scammed and the inevitable heartbreak of a poorly bred puppy is to arm yourself with knowledge.

Basic principles of supply and demand instruct that this epidemic of poorly bred dogs will shift only when people stop supporting puppy mills, USDA-registered mass-breeding facilities (factory-farmed dogs), and backyard breeders who don't focus on creating genetically healthy dogs. This means *never* buying a puppy on a whim; rather, view the process as on par with adopting a child: one that takes time, planning, and research. See page 403 of the appendix for a list of questions that a quality breeder will be able to answer before you choose to work with them). The questionnaire is so valuable because it gives you insights into epigenetic factors that can powerfully impact your dog's health down the road. For instance, if the pregnant mom is fed raw food and the puppies have early access to raw food, new research demonstrates a reduced likelihood of intestinal disorders and atopy (allergies). Unlike the environmental risk factors, over which we have tremendous control, our power over genetics is in working with reputable breeders and organizations that strive to improve the dog genome pool. Preservation breeders practice "reparative conformation" breeding, which means they have proactively completed all relevant DNA testing and health screens (and will happily show you the results) and are purposefully attempting to breed genetic flaws out of their animals. Functional breeders also work to diversify the gene pool, focusing on health, temperament, and purpose (function). These breeders understand that more must be done, beyond screening for the genetic diseases we know of and avoiding breeding dogs with known genetic issues.

Dr. Carol Beuchat of the Institute of Canine Biology elegantly explains why genetic testing and selective breeding alone won't fix our purebred genetic woes. In a nutshell, when a closed gene pool of dogs (purebred dogs all coming from the same family of ancestors) has litters of purebred puppies with no strategic genetic oversight (unintentional and sometimes intentional inbreeding), several bad things are more likely to occur: increased genetic similarity, increased expression of recessive mutations, decreased genetic diversity, and ultimately a decrease in the size of the gene pool.

As more and more purebred dogs are bred with each other, more genetic mishaps occur with bigger consequences, including shorter life spans. Geneticists call this "inbreeding depression." But there's even more to be depressed about: The risk of multigene disorders, such as cancer, epilepsy, immune system disorders, and heart, liver, and kidney disorders, also shoots up in these litters. But what about only breeding "the best of the best," the top 25 percent of champions that win at dog shows? Dr. Beuchat explains that by breeding only a tiny circle of purebred dogs we may be smashing our chances (by 75 percent) of ever finding enough unique and diverse genetic material to save some of our most beloved breeds. We're literally neutering our potential to identify these fuzzy genetic "diamonds in the ruff." In the long run, most people buying purebred puppies end up caring less about how popular or famous dad was and focus a whole lot more on how *healthy* dad was (or wasn't) once the vet bills start piling up. Her dismal conclusion: "There will be a steady deterioration in health of the population over generations unless there is appropriate intervention."

By intervention, she means replacing lost genes through "outcrossing" (crossing certain breeds to avoid certain genetic outcomes) to another dog population or introducing new genetic material via crossbreeding programs that will hopefully expand the gene pool of purebred dogs. Such methods are opposed by many breed purists. Every geneticist we spoke to reiterated this point: The way to improve the health of all dogs, long term, purebred or not, is proper genetic management. **Remember, genetic disorders result from the loss of the genes the body needs to function properly.** You can do everything right, but if your dog is missing the DNA for a healthy heart, he will develop heart disease. Create a mutation in the tumor-suppressor gene and cancer occurs. Take out the gene for healthy retinas and retinal dysplasia is the result. Remove the diverse genes of the immune system and immune disorders are inevitable. When animals carry genetic variants (SNPs), we have the potential to modulate their expression via epigenetics, but when genetic material is lost, there is no replacing it without expanding the gene pool (introducing new DNA), which means outcrossing.

DNA testing alone will not result in the creation of healthier dogs for years to come without a strategic, world-vision plan in place to intentionally reduce the consequences of selective breeding in a closed gene pool. This can be done only through thoughtful genetic management, exactly what the International Partnership for Dogs is attempting to do. And while DNA testing isn't going to ease the plight of purebred dogs, testing may be incredibly valuable to you, the dog health advocate you're in the process of becoming. Genetic testing can be an important step, as it pertains to identifying predispositions that can affect the long-term well-being of the pooch in your house. Our everyday habits deeply influence our genetic expression. While this is profound, what is truly life changing is that **we can change our lifestyle to improve that gene expression, thus increasing our chances of good health and longevity**. The same is true for our canine companions, with one caveat: It's up to us to make wise decisions for them.

Unfortunately, lots of breed DNA damage has already been done. Certain anxiety disorders, for example, cluster in specific breeds. In a 2020 Norwegian study looking at the connection between genetics and behavior, researchers found that Lagotto Romagnolos (a huge, furry retriever hailing from Italy), Wheaten Terriers, and mixed-breed dogs experience the most sensitivity to noise. Spanish Water dogs, Shetland Sheepdogs (Shelties), and mixed breeds tend to be the the most afraid. And while Labrador Retrievers rarely—if ever—show fear or aggression toward strangers, almost one-tenth of miniature Schnauzers display these traits. In a 2019 Finnish study, genes associated with sociability were also found to be on the same strand of DNA related to higher noise sensitivity. This shows that while humans have bred more social breeds, we have also created dogs for whom loud noises are an issue. Such trade-offs happen much more frequently than we probably realize. But with DNA research accelerating, we hope to limit the bad outcomes and push for better genetic management so we don't create more problems. It's not fair that we are preprogramming certain breeds to suffer from a panoply of ailments that could be prevented through proper genetic management. We are effectively killing some breeds entirely.

English Bulldogs, for instance, may have reached their genetic dead end. A breed known for short snouts and small, wrinkled bodies, English Bulldogs are so genetically similar to one another today that experts say it's impossible for breeders to make them healthier.

There are only two responsible options to acquire a dog in this world:

Option 1: If you are going to support a breeder, it's your duty to support only those breeders who are actively trying to improve their breeds' genetics. Use our twenty-point breeder questionnaire on page 403 to start the conversation with a prospective breeder. The website www.gooddog.com provides some good resources for discerning breeder quality.

Option 2: Adopt a dog from a reputable shelter or rescue organization. (These days, online puppy brokers pose as rescue or foster groups, so be aware. Pupquest.org has more information on this new scam.) When you decide to adopt a rescue or take in a homeless dog, you have no knowledge or say about the DNA your pup is carrying (if you're like us, this is a less important issue than saving the life in front of you). Many people simply refuse to buy a dog from a reputable breeder because their local shelter or breed-specific rescue is brimming with homeless dogs. More and more shelters and rescues complete DNA tests on litters of mixed-breed rescue puppies, recognizing that the more they know about the litter, the better odds they'll have for well-matched success. For example, adopting a puppy who is a mix of herding breeds means there's a strong likelihood he'll exhibit strong herding tendencies—good to know, prior to adopting! Rescuing is not for the faint of heart: Many people have endured repeated heartaches from indwelling issues with rescued pets. For instance, many shelter dogs are spayed or neutered at eight weeks of age; removing those key hormones before puberty can predispose many dogs to health and training challenges and lifelong hormonal imbalances that negatively impact the immune system years down the road. The choice to rescue or buy from well-researched heritage or functional breeders is a very personal decision. What's most important is to adopt *or* shop responsibly; extensive research is critical to understanding the massive, lifelong responsibility we assume when we bring a new furry friend home.

It's beyond the scope of this book to list every potential breed-related genetic flaw or mutation. You can get an idea of the recommended screening tests, by breed, at www.caninehealthinfo.org and www.dogwellnet.com. The best you can do as a pet parent is to determine the genetic makeup of the dog in front of you and, when possible, support his genetic deficiencies epigenetically by making wise lifestyle choices. Technology (i.e., DNA testing) has afforded us the power to influence our dog's genetic expression by positively influencing his environment and experiences. If you want to know what's under your dog's hood, test her DNA and then use the information in this book and at www.foreverdog.com to create a lifetime wellness plan that supports her unique genome. And if you don't want to know about your dog's specific markers, the science-backed suggestions in this book will go a long way to help you improve your dog's health span.

To recap, we can't alter our dog's DNA (or add back missing genes) but we *can* alter how his DNA expresses itself by influencing his epigenome through life choices (as a reminder, see page 88 for a list of epigenetic triggers; each is within your control). One recurring theme from many of the researchers we interviewed was the emerging science surrounding a dog's emotional health: we (humans) have long underestimated the power of our dog's social interactions to shape and influence physical well-being. Dogs are social animals and need social environments where they can develop social competence, express their personalities, and fully enjoy themselves.

Minimize Chronic Emotional Stress with Social Engagement and Stimulation

How many friends does your dog have? It shouldn't surprise you to learn that one of the three pillars of Blue Zone centenarians is strong social bonds, which translates into cultivating strong social networks for your dog. And don't underestimate the power of cuddles and kisses (if your dog likes close contact), your friendship is incredibly important to your dog; you may be his only social outlet.

To that end, we encourage you to constantly evaluate your job as a role model for your dog; address your own stress, and be as mindful, playful, and empathically communicative as possible. Building a deep abiding relationship with your dog is a lifelong process. Here's a tip: Once you've identified a few fresh food treats your dog loves, use them for short training sessions throughout the day or before and after work.

Even if you have an older, well-trained dog, it's still important to spend a few minutes each day working with your dog on communication skills. Dogs need jobs, or something interesting to think about, that engage their brains. If you don't want to spend a few minutes working on training or tricks each day, offer your dog a brain game or treat-release toy to focus on. And don't forget to make time to play, at least once a day. In her book *The Happiness Track*, Stanford University researcher Emma Seppälä notes humans are the *only* adult mammals that don't make time for play. Our dogs would *love* for us to engage in play more than we do—they're literally standing by waiting for us to interact with them. Play more . . . it's good for us, too.

Pro tip: Turning your phone on airplane mode when you're able to grab a few minutes of quality time with your dog is a great way to practice mindfulness and better connect with your dog.

It's no surprise that early life "exposures" and experiences set the tone for a dog's life. Studies show that the extent of appropriate socialization during puppyhood (between four weeks and four months of age) directly influences the dog's level of fear later in life (of both other dogs and strangers).

A dog's temperament is based largely on genetics *plus* the dog's experiences (or lack of experiences) in her first sixty-three days of life. For this reason, expert dog trainer and breeder Suzanne Clothier created her Enriched Puppy Protocol, which has been utilized to positively influence more than fifteen thousand puppies, including many destined to become service dogs. Dr. Lisa Radosta, a board-certified veterinary behaviorist, added that the dam's experiences and level of stress during pregnancy also play into a puppy's threshold for anxiety, fear, aggression, and phobias throughout life. Dr. Radosta said, "Those environmental circumstances lend themselves to how the puppy will behave later with the development of the brain and the temperament." Another important reason that in-depth conversations are imperative with a prospective breeder.

Dr. Gayle Watkins, from avidog.com, points out that breeder dogs in puppy mills live under constant environmental, emotional, and nutritional stress from churning out dozens of puppies— puppies who in turn are epigenetically influenced by their mother's stress and traumatic experiences, triggering the potential for all sorts of undesirable behavioral traits.

Developmental studies define three crucial socialization periods for puppies, with the first phase occurring at the breeder or rescue facility at four weeks of age. Puppies should begin early socialization programs at four weeks. These programs are designed to provide key sensory experiences during a very short window of time and are *invaluable* in building adaptable, easygoing temperaments in our dogs (see page 407 of the appendix for a list of early puppy socialization programs we love).

Assuming your puppy comes home around nine weeks of age, the next two crucial sensitive periods will occur with you. The next few months of your puppy's life are the most critical in laying the foundation for key behavior and personality traits, responses, and ability to cope with changes and variability in her environment for years to come. Appropriate and safe socialization primes your puppy with the coping skills she needs to navigate life. Well-socialized puppies grow up to be highly adaptable and have less cortisol, anxiety, fear, phobias, and aggression. By the same token, dogs who were not properly socialized as puppies are prone to higher stress responses (and cortisol) their whole lives.

Preventing fear of new situations starts at the breeder or rescue, when the puppies are four weeks of age. Puppies who are thoughtfully and safely exposed to the sights and sounds of the world on a daily basis from four weeks to four months (vacuums, gunshots or other loud noises, fireworks, storms, wheelchairs, kids, doorbells, to name a few) learn they don't need to panic or overreact to these events. These early life experiences either equip dogs to live confidently, abundantly, and adventurously or condemn dogs to live defensively, in active avoidance or defensiveness of new or unpredictable situations in a very scary world. Dr. Watkins emphasizes that the most important aspect about socialization isn't pushing a puppy into a scary world, it's about building and maintaining trust through new experiences that prepare our puppies for the fullness of life (within a short window of opportunity).

In-home early development programs and ongoing puppy classes help dogs start out on the right foot and aren't just highly recommended, they're *essential* if you want to create an emotionally resilient adult dog. In a nutshell, **exposures and experiences (good and bad) especially prior to four months of age can profoundly affect your dog's behavior and personality for the rest of her life.** This, in turn, affects her level of ongoing stress hormone production, which affects health span. Before you bring your new puppy home, take the time to curate a purposeful, diverse, engaging, and emotionally safe socialization plan.

Dr. Watkins stresses that relationship-centered, fear-free training classes should continue *at least* through your dog's first year of life. The juvenile and adolescent periods that occur from about six months through the first twelve to sixteen months can be challenging (the "teenage months"), and successfully navigating this trying period without aversive punishment is imperative for long-term psychological well-being. As Dr. Watkins said, "We need to remember that although they're physically grown and look like adults, they are still very much cognitively developing." Unfortunately, undersocialized "pandemic puppies" are now emerging in the world as unruly, reactive teenagers, causing a lot of parental stress. What's most important is that you make a plan, right now, for rectifying the situation (with the help of trained professionals, using science-based, humane training methods). "If you make a great puppy, you have a great dog."

"TEACH ME HOW TO BE THE DOG YOU WANT ME TO BE."

Our belief mirrors that of animal behaviorists: ongoing, lifelong, relationship-centered training isn't an option, it's an obligation. It's not something you start once your puppy or rescue pup develops an annoying behavior, it's how to *prevent* behavior problems in the first place.

It's never too late to introduce your dog to new experiences, as long as you go at a pace that doesn't create anxiety or fear. "Learning how to read your dog's body language is the most important thing you can do," says Dr. Radosta. Being able to accurately read your dog's nonverbal communication is critical for lots of reasons, including early intervention when there's excessive stress associated with negative experiences (check out *Doggie Language: A Dog Lover's Guide to Understanding Your Best Friend* by Lili Chin for a good first read on canine body language). Pupquest.org reports that up to 50 percent of puppies don't last a full year in their first home, and only one in ten dogs spends his or her whole life with the same family. Re-homed animals can show signs of PTSD, among myriad other behaviors often requiring professional intervention for the most successful outcomes. If a puppy hasn't been adequately socialized, you can commit to damage control (behavioral modification) at any age to help improve your dog's sense of security and happiness; depending on how reactive or shut down your dog is, the endeavor may require professional help. We recommend getting well-credentialed help sooner than later to manage any recurrent behavioral issues or concerns. The sooner you address the issues, the sooner things will improve. Choose your dog trainers wisely, like you'd choose a nanny for your child. See our list of suggestions in the appendix on page 407.

In addition to equipping your dog with the social and emotional skills to be happy, functional, and relational in your home and community (or sensitively managing those who cannot), identifying and addressing sources of repetitive stress is also a wise idea for curbing the potential for lifelong anxiety. Veterinary visits, nail trims, ear cleanings, and baths are just a few of the common experiences that may cause your dog to become uneasy. **Learning how to appropriately manage your dog's stress responses is one of the best assets in your relationship toolbox and a lifetime gift to our dogs.**

Our friend Susan Garrett's forte is training world-class canine athletes. She's best known for being a ten-time world champion in dog agility, but she's also exceptional at solving everyday challenges we all experience when attempting to communicate with another species. She reminds us that if you own a dog, then you're a dog trainer by default, and good dog training is nothing more than growing two critical elements: your dog's self-confidence and her trust in you. You can accomplish both goals simultaneously by approaching every interaction with your dog believing she will always do the best she can with the education you've given her. Dogs never want to let us down. Unfortunately, dogs are routinely scolded for behaving like "a dog"; consequently, their trust in their people gets routinely broken. As we've mentioned, your relationship with your dog is built on trust and excellent two-way communication. Whether you have a rescue dog or a new puppy, daily education (training) is required to grow and maintain your dog's understanding.

When dogs experience stress or fear (fireworks, strangers at the door, a beeping smoke detector, a new harness, car rides, the vacuum, and so on), they respond reflexively, without conscious decision making; their bodies are wired for self-protection. As their guardians, what's most important for us to remember is that **stress and fear are instant roadblocks to learning**, Susan points out. It's impossible for man or beast to "learn" when a fear response is triggered. Stress hormones are instantaneously released and prompt a fight, flight, or freeze response, offering a primal way for a dog to protect himself against the perceived threat. The body immediately allocates all of its resources to "survival mode," and for dogs, the fear response culminates in growling, snapping, barking, lunging, cowering, panicking, and/or escape behavior.

During stressful situations, it's unlikely our dogs will respond to our words as they normally would, unless we've trained them to have an alternative, healthier response—a coping mechanism during intensely stressful situations. They can't hear us because they're in panic mode. Don't punish your dog for panicking. Instead make it your goal, with the help of professionals if needed, to create a positive "conditioned emotional response" when they show signs of stress or fear. And although we can't "train" our dog in the moment of stressful situations, we can begin to condition our dog to experience the stressful situation differently. **We have the ability to help our dogs succeed at coping with stressful situations and get through potentially fearful situations by deepening their trust in us, rather than eroding it.**

With a commitment to the process, we can change the triggers so that in the future, anytime your dog encounters the previous trigger, they can respond in a more desirable way, looking to us for reassurance and rewards, rather than being triggered in fear.

Fearfreepets.com is a great resource if you are looking for veterinarians and groomers that make it their mission to prevent and alleviate fear, anxiety, and stress in pets by inspiring and educating the people who care for them. Taking the "pet" out of "petrified" is their tagline. The point is, do everything you can to help your dog overcome emotional obstacles that cause a recurrent stress hormone secretion that is unhealthy for her body. And do everything you can to try to keep your dog in a state of emotional balance . . . the same goes for you.

Of course, you aren't going to alleviate all the stress in your dog's life; the world is a crazy place full of unpredictable, scary events. But managing the known, day-to-day or repetitive stressors *is* within our reach, and we owe it to our dogs to begin the laborious but very rewarding process of desensitization and counterconditioning (behavior modification techniques your trainer will use) so that next year will be less stressful than this year. If you do nothing (except react), unwanted behaviors will likely get worse. And so will your relationship.

Our goal is to be trustworthy, consistent, and reliable with our responses to those around us; this is crucial for not inadvertently creating crazy dogs. I (Karen) learned soon after adopting Homer that touching his feet were off limits (or risk being bit), and baths were an apparent near-death experience (or risk a senior citizen panic attack). Six months after adoption (a remarkably short time), I'm proud to say Homer will eat treats while standing in a foot bath, unrestrained. Commit to implementing science-based "damage control therapy" to address unwanted behaviors in your dog. Do it, because the alternative isn't good or healthy for either of you.

Dog Days (Decided by Dogs)

If we let our dog choose what they want to do during a day, what would they choose? Looking at life from our dogs' perspective is something we should do more often. What activities are exciting to them? What foods are most enjoyable? What do they want to smell, and who do they want to interact with? Learning about our dogs' preferences makes us better guardians and enhances our bond, not to mention their quality of life. The more often we set aside time to get to know our dogs' preferences, the more competent and capable we'll be of meeting their social, physical, and emotional needs.

Don't take your dog to dog parks unless you know how your dog will respond, or it will exacerbate your dog's stress (and yours). Suzanne Clothier made this point clear when we spoke with her: **Dog parks are the worst choice for undersocialized or shy dogs**. If you want to create positive, outdoor experiences for your reactive or fearful dog, you'll have to commit the time it takes to repattern your dog's behavior—at a pace and with training techniques that don't stress him out (research is clear that punishment-based training exacerbates anxiety and further increases stress hormones). Many of us have rescued poorly socialized dogs with some emotional baggage and erroneously assumed that a loving, stable environment will fix their mental and emotional issues. "It won't," says Dr. Radosta. When you adopt or rescue a dog that exhibits behavior issues (including fear and anxiety), all the love in the world won't fix them; you need to address the issues immediately, and preferably with a team of professionals: "Assemble your behavior modification team as if you were planning your wedding," she counsels. The American College of Veterinary Behaviorists has a directory at www.dacvb.org.

The most important thing we can do for our pups is to find and provide safe experiences, activities, and exercises *they* really enjoy, based on their personality and physical abilities. Dogs have preferences, just like us, and discovering our dogs' joys in life will feed our souls. If you don't know what your dog likes to do, make it a point to try lots of things. Even activities your dog didn't respond to early in life can be much more interesting midlife or as seniors, so explore away!

Let's also not forget the effects of ongoing positive mental stimulation on the brain. Earlier we detailed studies that show how the combination of an anti-inflammatory diet with social experiences and proper exercise promotes higher levels of a very important growth factor—BDNF—in the brain. This is your brain's way of nourishing its cells and fostering the birth of new brain cells—a good thing at every age!

Veterinary behaviorist Dr. Ian Dunbar believes one of the best things we can do for our dogs' emotional well-being is to **cultivate a rich social life for them**: identify a handful of dogs your dog really likes (dog friends) and then make it a point to have playdates throughout your dog's life. Dogs need *lots* of regular opportunities to be dogs: to sprint at full speed, dig in dirt, roll around on the ground, smell butts, play, tug, gnaw, bark, and chase. *You* provide these opportunities. Another job title we have as our dog's care-taker is that of a boredom buster. Many well-loved dogs live gener-ally boring lives, and not by their own choice; rather, they aren't in control of the lives they are living.

Julie Moriss told us she regularly scheduled playdates for Tig-ger, her twenty-two-year-old Pit Bull, especially as she aged, so she could have social interactions with her dog friends. As trivial as it sounds, Dr. Kubinyi's research confirms this may be as emotionally important for dogs as Blue Zone research confirms it is for humans. We are both social species and need ongoing opportunities for pos-itive, social engagement throughout our lives.

If your dog doesn't have the social skills to be in groups with other dogs, then find something she really enjoys doing with her brain and/or body and do it regularly. Nose work (scent games) is our favorite activity (turned into a dog hobby, or a "job," for work-ing breeds) for aggressive, reactive, and shy dogs as well as for dogs with PTSD. Dr. Radosta says it's our responsibility as guardians to provide our dogs with the "five freedoms":

- ➤ freedom from distress (fear/anxiety)
- ➤ freedom from pain or injury
- ➤ freedom from environmental stress and discomfort
- ➤ freedom from hunger and thirst

➤ freedom to express behaviors that promote well-being and to engage in species-typical behaviors

Dr. David Mellor, professor of animal welfare science at Massey University in New Zealand, went one step further and developed a set of guidelines he calls the Five Domains. His model emphasizes maximizing positive experiences, and not just minimizing negative ones, which can have longevity enhancing benefits:

➤ Good nutrition: Provide a diet to maintain full health and vigor and enable an eating experience that is pleasurable.
➤ Good environment: Minimize exposures to health-insulting chemicals.
➤ Good health: Prevent or rapidly diagnose/treat injury, disease, and illness. Maintain good muscle tone and physical function.
➤ Appropriate behavior: Provide congenial company and variety, minimize threats and unpleasant restrictions on behavior, and promote engagement and rewarding activities.
➤ Positive mental experiences: Provide safe, enjoyable species-specific opportunities for pleasurable experiences. Promote various forms of comfort, pleasure, interest, confidence, and a sense of control.

Providing excellent nutrition and a low-stress, nontoxic living environment; maintaining a healthy physical body; engaging in rewarding activities; and creating positive mental experiences—Blue Zone researchers endorse each of these methods to yield a robust, long life.

Last but not least, observe and listen to your dog. Pay close attention to everything—body, body language, and behavior. Get to know your dog as well as you know your other kids or the person closest to you on this planet. Learn to know when your dog is uneasy, learn to know his preferences—his favorite time and way to play, where and how he likes being touched, what he enjoys doing, what foods he *really* likes. When you make it a point to make your dog your best friend, or at least a valued member of your family, you'll become a better guardian (and dramatically improve the quality of your dog's life and your relationship).

You will watch more, you will engage differently, you'll be more sensitive and more connected, and you'll ask yourself better questions: Why has she licked the top of her right paw two evenings in a row? Your thought process will expand beyond "My dog barfs on the rug after he licks the carpet" to "Why does he want to lick the carpet so much?" You'll be compelled to dig for the root causes of their distress. You'll begin to look at your dog's behaviors and choices as a day-to-day road map for what you, as their advocate, need to address or do. We begin seeking to understand our dog's behavior rather than reacting to it. In this way, we do our part in maintaining our commitment to do the very best we can for the animals relying on us. We can't let them down. But to do it right, we need to know our dogs very well. To keep them healthy in our homes, we need to take a deeper look at our dogs' immediate environment.

Minimize Environmental Stress and Reduce the Chemical Load

Chapter 6 may have had you running for the hills, recognizing just how toxic our modern lives are with all that surrounds us and everyday exposures. From the moment we get out of bed, which itself could be filled with off-gassing chemicals, we encounter countless sources of environmental toxins. Some are relatively harmless, and some are inevitable, such as veterinarian-prescribed pesticides for fleas and ticks and heartworm prevention. Some veterinary chemicals are important to prevent disease, but all still require your dog's body to metabolize and excrete them. I (Dr. Becker) see many dogs' liver enzymes elevate over the summer and return to normal in the

winter, when prescribed chemical pesticide application and inges-
tion diminishes. Veterinary-prescribed chemical exposures can add
to the overall chemical load—that body burden—and increase risk
factors for disease. How well did you score on the chemical quiz on
pages 179–80?

Don't panic if you feel "toxic"! The point of highlighting your
exposures is to empower you to make changes that will go a long
way to protect you and your pooch and limit future exposures. The
goal is to prevent exposures from disrupting the essential functioning of the body, impacting DNA, cellular membranes, and protein.
Below is our thirteen-point checklist to follow in cleaning up your
environment. We realize that some of these strategies have been
mentioned or implied in previous chapters, but it's helpful to have a
list all in one place, so **start here:**

1. **It starts with food:** Minimize metabolically stressful food
   that prompts cortisol and insulin spikes (get that starch out!).
   If you've already implemented the strategies in the previous
   chapters, you're on the right track. Fresher food diets also
   minimize your dog's consumption of harmful mycotoxins,
   food chemicals and residues, and by-products of high-heat
   processing (AGEs).

2. **Remove plastic water bowls,** as they're full of phthalates
   that disrupt the endocrine system. Instead, use quality stainless steel, porcelain, or glass. In the steel department, choose
   eighteen-gauge stainless steel and preferably from a company
   that conducts third-party purity testing because even stainless
   steel has proven to be contaminated (remember the Petco metal
   bowl recall several years ago?). In the porcelain department,
   be aware that some porcelains can contain lead and others are
   not approved for food products, so make sure you buy good-
   quality porcelain made for food use from a company you trust.
   Pyrex or Duralex glass bowls are our favorites, as they're durable and nontoxic, unlike other cheaply made glass products that

may contain lead or cadmium. Note, too, that there is a general tendency among many pet parents to buy oversize food bowls for their dog. People typically fill their dog's bowl to the brim since a half-empty bowl just doesn't look sufficient for a hungry pup. This is a mistake; portions should be based on your dog's caloric needs, not their dinnerware. If the bowl you have is too large for your pet, you could convert it to a fresh water bowl. Many pet parents have a massive food dish and a tiny water bowl, when in fact the opposite may be best. Water is vital for dogs.

3. **Filter your dog's water.** Regardless of how much you love the taste of your tap water or the glowing report from your water supplier about the contents of your water, buy a household water filter, at least for all of your drinking and cooking purposes. The chemicals we produce and use in industry and agriculture eventually make their way back into our drinking water. A household water filter effectively removes a large amount of toxins your dog would otherwise consume from city or well water. A variety of water treatment technologies are available today, from simple and inexpensive water filtration pitchers you fill manually, to under-the-sink filtration systems with storage tanks, to whole-house carbon filters that filter all of the water coming into your home from its source. The latter is ideal, particularly if you subscribe to a service that changes the filters regularly, because you can then generally trust the water used in your kitchen as well as the water in your bathrooms. Choose the filter technology that best suits your circumstances and budget: whole-house carbon; individual carbon filters on water spigots, refrigerators, and the like; reverse osmosis filters in the kitchen. Do your own research, as each type of filter has its strengths and limitations, and one type does not accomplish all goals.

4. **Purge the plastic:** Minimize the amount of plastic you use in your life. It's impossible to avoid plastic, but you can certainly

limit its presence and thereby your (and your dog's) exposure to more phthalates and BPA. Use common sense when deciding how to store your dog food and your own beverages. Use quality glass, ceramic, or stainless steel whenever possible, and avoid storing food in plastic bags. Never, ever microwave, cook, or bake using any plastic. When buying toys for your dog, skip the plastic toys and look instead for toys labeled "BPA Free" or those made in the United States from 100 percent natural rubber, organic cotton, hemp, or other natural fibers.

5. **Remove shoes when entering your home and wipe those paws:** In many countries, it is customary to remove your shoes upon entering a house. It shows respect for the home and its occupants. However, in many Western countries, including the United States, it is uncommon to leave your shoes at the door (or outside). But leaving shoes outside is one of the easiest things you can do to avoid exposures to harmful substances, ranging from pathogenic bacteria, viruses, and fecal matter to toxic chemicals, including a whole panoply of chemicals we'd all do well to avoid. Your shoes carry contaminated dust from nearby construction sites as well as chemicals recently sprayed on lawns, around the perimeters of houses, near public parks, and even on the sidewalks outside your home. Because dogs are naturally closer to the ground, this strategy is extra important. You can take it one step further by wiping your dog's paws with a wet cloth (use castile soap, if needed). For folks who live in cold climates where the roads are salted in the winter, this is especially important. Winter road salt sickens lots of dogs: use "pet friendly" salt or sand at home.

6. **Clean up your air: minimize sources of volatile organic compounds (VOCs) and other noxious chemicals:** Get yourself a good vacuum with a HEPA filter if you don't already have one. HEPA stands for "high-efficiency particulate air." To qualify as a HEPA filter, the product must remove 99.97 percent of airborne particles measuring 0.3 microns or greater in diam-

eter. To put 0.3 microns in perspective, a typical human hair ranges from 17 to 181 microns in diameter. A HEPA filter traps particles several hundred times thinner, including most dust, bacteria, and mold spores. VOCs often adhere to dust, so a HEPA vacuum will help you minimize flame retardants, phthalates, and other VOCs in your household. Watch out for VOC-laden air fresheners, candles, plug-ins, and carpet cleaners. We suggest that you simply ban all air fresheners, sprays, plug-ins, or scented wicks from your home. These are laced with phthalates and myriad other chemicals. ***When in doubt, take it out!*** If you have carpets, try to vacuum thoroughly and as often as you can (at least once a week). You also can add HEPA air purifiers in the rooms where you spend the most time (living room, den, bedrooms, and so forth). Use exhaust fans wherever you have them, such as the kitchen (while cooking), bathroom (while bathing, showering, or spraying personal-care products), and laundry areas (while doing the laundry). Wipe windowsills with a damp cloth and vacuum blinds regularly. Damp-mop tile and vinyl floors and vacuum or dust-mop wood floors regularly—weekly, if possible. Keep any toxic materials that you feel are necessary, such as glues, paints, solvents, and cleaners, in a shed or garage—away from your living quarters.

7. **Rethink outdoor lawn care:** Outdoor lawn service chemicals, including fertilizers, pesticides, and herbicides, are much more toxic for dogs because they don't wear protective clothing or shoes and they don't shower regularly to remove accumulated chemicals. Natural pest control services and lawn care are available. Replace Roundup and other synthetic pesticides or herbicides in your garden shed or garage. A variety of less-toxic organic herbicides are available that effectively eliminate weeds with safer options (www.avengerorganics.com offers products popular with pet lovers) and without increasing your family's risk of cancer. Chemical-free lawn maintenance

programs such as www.getsunday.com are popping up around the globe, delivering right to your door ready-to-apply, chemical-free, all-in-one lawn maintenance kits (tailored to your soil, climate, and lawn).

Try to find ingredients free of synthetic agrochemicals, and instead look for herbicides with citric acid, clove oil, cinnamon oil, lemongrass oil, d-limonene (from limes), and acetic acid (vinegar). One natural herbicide, corn gluten meal, is a common ingredient in dry dog food but is better suited as a remedy for crabgrass. Don't forget the use of beneficial nematodes you can release into the garden or yard that feast on flea larvae, ticks, aphids, mites, and other bugs and are harmless to people, plants, and pets; the www.gardensalive.com website is a good place to start learning. Replace your traditional garden hose (that leaches lead, BPA, and phthalate) with an NSF (National Sanitation Foundation) -certified, phthalate-free potable water hose. If you can get PVC-free, so much the better. And use that hose to rinse off a dog that's recently come out of a swimming pool! Have a look at the Sustainable Food Trust website (www.sustainablefoodtrust.org) for more information and ideas about sustainable and organic gardening.

VETERINARY PESTICIDE SUPPORT

When it comes to how often and to what extent you should use flea, tick, and heartworm pesticides, you can make common-sense choices, like assessing risk versus benefit. If your Maltese rarely leaves the backyard, which is regularly serviced for pest control, your dog's relative risk of acquiring a massive tick infestation is significantly lower than if you often camp and hike with your dog deep in the woods. If you frequently spend time in the woods with your dog in high-risk areas, then you will need to provide protective chemicals and support your dog's endogenous detoxification pathways (see the protocol laid out in Chapter 4).

"Deterrents" or natural repellents (typically made with botanicals or less-toxic chemicals) make your dog less attractive to parasites but aren't always 100 percent effective (and neither are chemical pesticides, FYI). "Preventives" are chemicals (pesticides) that have been approved by the FDA for application on or in dogs. Each chemical is approved for killing specific parasites or several types of parasites. These veterinary pesticides present a broad range of potential side effects. In 2003, the USDA granted organic status to Spinosad, an environmentally friendly insecticide. It is a relatively new insect killer derived from the fermentation juices of a soil bacterium called *Saccharopolyspora spinosa*, so it's toxic to pesky insects but not to mammals and may be a safer option than isoxazoline products (Bravecto, Simparica, and Nex-Gard). In a recent survey of dog owners using isoxazoline products, *66 percent reported some kind of a reaction to the ingredient.* On September 20, 2018, the FDA issued a warning that products containing isoxazoline cause adverse events in pets, including muscle tremors, ataxia, and seizures. The FDA worked with manufacturers of isoxazoline products to include appropriate neurologic warning on their labels.

Every pesticide comes with its own set of risks and benefits, depending on how well your dog's detox pathways are functioning (which dictates how well he can clear chemicals from his body), dosing frequency, immune status, and other variables.

Every dog should be *individually assessed* based on their own unique risk profile. Remember: Many dog parasites, like ticks, harbor diseases that also can infect humans, so you are just as much at risk as your dog. We recommend doing for your dogs what you do for your outdoor-loving kids or yourself, in terms of choosing parallel pesticide protocols. To determine an appropriate parasite control program for your dog, consider the following:

- ➤ Does my dog have any underlying medical issues that would complicate the clearing of pesticides from his body (liver shunts, abnormal liver enzymes, or other congenital birth defects)?
- ➤ Do I live in a low-, medium-, or high-risk area for certain parasites?
- ➤ If I live in a medium- or high-risk area, how often are we exposed: daily, weekly, monthly?
- ➤ Is our exposure year-round?
- ➤ Am I willing to regularly and thoroughly check for visible external parasites (like fleas and ticks) on myself and my dog? This is an important question, as it's the primary way to identify the creepy crawlies one of you may have picked up outside.
- ➤ Do I have a detox protocol ready to go? If you live in a very high-risk area and spend a lot of time outside, you're probably stuck using some chemicals, but the type and frequency of applications should be adjusted during lower-risk months. If you use chemicals, we recommend a detox protocol, because your dog's pesticide body burden will be unrelenting. The microbiologists we interviewed suggested using probiotics and microbiome-supportive protocols if flea and tick chemicals are routinely used.

If you live in a high-risk environment but have low exposure, or a low-risk environment and have high exposure, you may find a hybrid parasite protocol makes the most sense: rotating natural deterrents and chemical preventives. In any tick-endemic area, we recommend performing a tick-borne illness screening test with your veterinarian at least annually, regardless of what prevention strategy you use. See page 398 of the appendix for details.

DIY PEST SPRITZ

1 teaspoon (5 ml) neem oil (from the health food store or your favorite high-quality essential oil manufacturer)

1 teaspoon (5 ml) vanilla extract (from your kitchen cupboard; this helps the neem oil last longer)

1 cup (237 ml) witch hazel (which helps the neem oil disperse in the solution)

¼ cup (60 ml) aloe vera gel (to help keep the mixture from separating)

Add all ingredients to a spray bottle and shake vigorously until mixed well. Immediately spritz over dog (avoid eyes!). Repeat every four hours while outdoors. Shake well prior to each use. Always flea comb your dog after being outside to remove any creepy crawlies (remember, no pesticides or natural deterrents are 100 percent effective). For maximum potency, make a fresh batch every two weeks.

DIY PEST COLLARS

10 drops lemon eucalyptus oil (purchase all these oils from your favorite high-quality essential oil manufacturer)

10 drops geranium oil

5 drops lavender oil

5 drops cedar oil

Mix oils together and apply five drops to a cloth bandanna (or cloth collar); let your dog wear the bandanna while outdoors. Remove bandanna after your outdoor excursion. Reapply five more drops to bandanna daily, before you spend time outside. Again, always flea comb your dog after being outside to remove any creepy crawlies.

Note: Don't use these products if your dog has sensitivities to any of these ingredients.

Practice the precautionary principle: The precautionary principle holds that when the effects of a chemical are unknown or disputed, minimize exposure now to avoid suffering the consequences later. When it doubt, take it out!

8. **Rethink general household goods:** Invest in an organic dog bed that is made with wholly natural materials. Unless you are financially prepared to buy a new, organic mattress for yourself (most dogs end up on our beds), the best you can do is purchase a barrier cover made of 100 percent organic cotton, hemp, silk, or wool. You can do this for your dog's bed, too, if you're not prepared to buy him a new organic dog bed; a simple organic cotton sheet or throw blanket works. Wash it weekly with VOC-free detergent, and don't use fabric softeners. When you buy household detergents, disinfectants, stain removers, and so on, select green cleaning products with simple ingredients that have been around for eons (for instance, white vinegar, borax, hydrogen peroxide, lemon juice, baking soda, castile soap). The escalating use of chemical-based consumer products inside our homes is disastrous for dogs, especially those who spend most of their time inside. Carefully evaluate any products you bring into your home. Be wary of labels that say "safe," "nontoxic," "green," or "natu-

ral," because these terms have no legal meaning. Read labels carefully, identify the ingredients, and pay special attention to warnings. You can accomplish a lot using harmless, effective, and economical ingredients to make your own cleaning products. Thousands of easy recipes online use well-known, nontoxic ingredients. Remember the "fragrance exception": Under federal law, manufacturers are not required to disclose the chemicals of any substance labeled "fragrance," a not-so-nifty loophole disingenuous companies exploit to mask toxic ingredients. If you must use chemicals in your home that are labeled as caustic or carry a "call poison control if ingested" warning, do a second and third pass with pure water to remove all trace chemical residues.

9. **Consider dog hygiene:** Choose organic or chemical-free grooming products, from dog shampoos to ear cleaners and toothpastes. Evaluate the ingredients in the products you put in or on your dog; for instance, many powders that remove tearstains are low-dose antibiotics (Tylosin) that can disrupt the microbiome over time; and most deterrents for coprophagia (poop eating) contain MSG (monosodium glutamate), which can cause behavioral disturbances and neuroendocrine issues in animal models.

> **Homemade toothpaste recipe:** 2 tablespoons baking soda + 2 tablespoons coconut oil + 1 drop of peppermint essential oil (optional); combine ingredients, mix well, and store paste in glass jar. Wrap finger in gauze, dip gauze-covered finger into paste, and massage mixture on your dog's teeth nightly, after dinner.

10. **Maintain oral health:** We all grossly underestimate the power of oral hygiene. But the science is clear: Oral health plays into everything about us, including how much systemic

inflammation we bear. When our mouths and gums are clean and free of infection, we reduce our risk for dangerous inflammation and dental disease. It's estimated up to 90 percent of dogs have some form of periodontal disease at one year of age. Many human toothpastes contain xylitol, a sweetener that can be lethal to dogs. Fluoride is also not safe for your dog, so use oral hygiene products made specifically for pets. You can further support your dog's oral health with raw bones. One study from Australia found that 90 percent of dental calculi were removed within three days just by offering raw, meaty bones! (See page 405 in appendices for Raw Bone Rules.)

11. **Opt for vaccine antibody titer testing:** A titer test is a simple blood test that gives pet parents an understanding of a pet's resistance to certain disease based on past vaccinations. Adult humans don't receive annual booster shots for the core vaccines we received as kids because our immunity lasts for decades, and in most cases for life. Similarly, a puppy's core vaccine regimen typically provides immunity that lasts for many years (and often a lifetime). As opposed to following the legal requirements for automatic revaccination many countries enforce, consider titer testing. It can also help lower your pet's chemical (adjuvant) load by providing only what is needed—without going overboard—to maintain an alert immune system. No further vaccines are required if the titer test comes back with a positive antibody, which indicates your pet has an effective immune response. All of the Forever Dogs we met had modified vaccine schedules; they were vaccinated as puppies but not annually, as adults.

12. **Keep noise and light pollution in check:** Bring as much natural light as possible into all the rooms of your house so you don't have to resort to using so much artificial lighting. Both fluorescent and incandescent lighting lack the full spectrum of wavelengths that exist in sunlight. Depriving your dog of

natural sunlight has known health consequences, from chaotic circadian rhythms to depression. We need to do a better job of honoring our biologic clocks. Dim lights in the house after dinner or by eight p.m., at the latest, and minimize or shut off any blue-light-emitting screens (phones, computers, etc.). Dr. Panda has no overhead lights on in his house after dinner. We have followed suit. He has a good, memorable saying: "Light for vision is not the same as light for health." You can buy cheap tabletop dimmers for under $10 for your lamps, and dial back the brightness closer to bedtime to keep melatonin levels balanced and turned on. Keep a quiet room in the house and manage external sources of extraneous, ear-jarring noise such as loud TVs. Turn off your router at night. Create a safe haven for your dog that's dark, cool, and calm if you're the type who stays up late with entertainment (i.e., more light and noise) and it's past your dog's bedtime.

13. **Build a proactive, wellness team:** Small, independent micro-retailers in the pet space are your best bet when it comes to supporting you on your dog's wellness journey. These locally

owned retail shops typically are staffed by wildly passionate dog lovers who are well educated about pet food choices and have done their research on the brands they carry in their stores. They also are well connected with other professionals in the animal wellness community so they can point you in the right direction when you're looking for rehab/physio help, a local trainer, or a proactive veterinarian. In many instances, just like for human health, you end up choosing your dog's health-care team like you do your own. Many of us have a family doctor or a GP, an ob/gyn, a chiropractor or massage therapist, a nutrition counselor, a personal trainer, a dentist, a dermatologist, a therapist, and a podiatrist in our lives for maintenance, health, and disease prevention. As we age, this list grows to include oncologists, cardiologists, internists, surgeons, and more. The goal is to à la carte our own well-being so successfully— blending the expertise of many practitioners who have a unique focus on one part of the body or one aspect of healing—that we don't require specialists later because we prevented illness through wise lifestyle choices. The days of one country doctor doing everything for us, medically and physically speaking, are gone. Health diversification has reached veterinary medicine in many parts of the world, as well. Many pet owners have a regular vet, an integrative or functional medicine wellness vet, an ER clinic for after-hours care, a physiotherapist to rehab from injuries (or prevent them), and an acupuncturist and/or chiropractor. If you live in a rural area or otherwise don't have access to a smorgasbord of wellness services, don't fret: Reading this book is an excellent start, and the Internet is full of credible resources to help you become an informed, savvy health-care advocate for yourself and your dog.

Finally, there are many changes you can make *for free* to improve well-being. You do not have to be wealthy or wait for that annual bonus to incorporate the ideas in this book.

Exercise provides or replaces the need for more than two dozen supplements and is a natural way to detoxify your dog's body. *Free detox!* If you are short on cash, use exercise as a powerful antiaging tool. We must give our dogs daily opportunities to *move* for them to make adequate BDNF, which doesn't come in supplement form. BDNF is lowered by stress and free radicals, and it's boosted by aerobic exercise and adequate vitamin B5 (found in mushrooms!).

Don't forget about those melanopsin-releasing morning walks and melatonin-secreting evening walks. Open your blinds and maximize natural sunlight, turn off your router at night, engage in daily DIY brain games, schedule playdates, maintain your dog's weight, practice time-restricted feeding, and stop feeding your dog at least two hours before bed; creating an eating window alone can bring about profound improvements in your dog's metabolic and immunologic well-being.

These are just a few suggestions covered in this book that will stack the deck mightily in your dog's longevity favor. It doesn't take much to get economically resourceful. When buying your own groceries, for instance, seek out dented and bruised produce on sale and freeze it. Recycle leftover veggies (without sauces) into your dog's bowl. Grow your own veggies. Join a food co-op. Get to know the people at your local farmers' market. Save the parsley garnish from your takeout as a CLT freebie, top off your dog's food with the safe culinary spices in your spice drawer, boil up your leftover chicken bones and make broth, make herbal tea for two (cool before serving over your dog's food), take your dog to the woods to play in the dirt, swim in the lake. The sky is the limit for innovative and economical day-to-day choices that tip the scale toward a longer, healthier life for your dog!

The process of incrementally improving our dog's health and well-being is a journey, an evolution of sorts. Dogs and humans have co-evolved for millennia, relying on each other, learning from each other, listening to each other, and symbiotically enhancing each other's physical and emotional well-being. As you embark on the adventure of creating your Forever Dog, remember: Dogs live in the moment, in the now. Now is the best opportunity we have to enhance our health journey together, to have the longest and most fulfilling "walk home."

LONGEVITY JUNKIE TAKEAWAYS

➤ Dogs should get a bare-bones minimum of twenty minutes of sustained heart-thumping exercise at least three times a week; most dogs benefit from longer, more frequent sessions. Try swapping twenty minutes for thirty minutes to an hour, and change three days a week to six or seven. Get creative with exercise, and find out what your dog likes to do.

➤ Brain exercises are as important as physical exercises (www .foreverdog.com has more suggestions).

➤ In addition to daily exercise, take your dog on circadian-setting sniffaris twice daily—morning and night. At least once a day, let her sniff whatever interests her for as long as she wants; no leash yanking.

➤ Whether you adopt or shop for your dog, do so responsibly. If you buy a dog, partner with a qualified, well-researched breeder who focuses on creating genetically healthy puppies (see page 403 of the appendix for the questions you need to ask prior to purchasing a dog).

➤ If you rescue, it's okay not to know the DNA of your dog; if you're curious, you can explore genetic testing (see www.caninehealthinfo.org and www.dogwellnet.com). Many genetic predispositions can be positively influenced epigenetically by lifestyle factors you control.

➤ Minimize chronic emotional stress by providing ongoing social engagement, mental stimulation, and dog-centric activities your dog loves.

➤ Minimize environmental stress and reduce your dog's chemical load with the thirteen-point checklist.

Let's look at a sample day in the life of a Forever Dog. The vast majority of you reading this book don't live on a farm, where you can open the door and let your dog go live his best life, experiencing the day on his own terms. Most of our dogs are waiting for *us*. The responsibility falls on us to create meaningful opportunities for choice, exercise, engagement, and play. Make the commitment to intentionally create "good dog days" that provide experiences that nourish your dog's body and brain.

What does a good dog day look like? It looks different for everyone, but we've been fortunate to see how thousands of people in our community have put Forever Dog principles into action, cultivating Forever Dog days that work for their lifestyle. Here's one from Stacey and Charm.

Stacey is a twenty-six-year-old professional dog walker from Pittsburgh. Charm is an eight-year-old rescued Yorki-poo. On days Stacey starts work really early, there isn't time for morning exercise, so her routine looks like this:

- All house window shades/blinds open first thing in the morning.
- Fresh, filtered water in a Pyrex glass bowl.
- Dog breakfast: a small amount of homemade food mixed with a tablespoon of synthetic-free kibble, topped with warm bone broth, supplements hidden in food (she doesn't feed a big meal in the morning so Charm won't have to poop while she's working).
- Ten-minute circadian-setting sniffari (opportunity for Charm to sniff, poop, and pee and sniff some more, and for Stacey to drink a cup of coffee while getting some fresh air).
- Stacey goes to work for six hours, comes home for a late lunch. Uses CLTs for training treats while heating up her own lunch, works on "sit," "stay," and "down" for a few minutes. Loads an interactive brain game toy with freeze-dried meat for Charm to work on while she's eating lunch.
- Twenty-minute power walk with Charm, then back to work.

- Stacey gets home from work, plays tug as a warm-up, then rigorous ball chasing in the backyard. Dog dinner (the bulk of Charm's calories): freeze-dried food reconstituted with Medicinal Mushroom Broth (see page 225), plus wellness supplements. Stacey minces the same veggies she's making for her dinner and mixes them into Charm's meal for her CLTs.
- Uses a tablespoon of commercial raw food on a lick mat (a great distraction tool) to engage Charm while she eats.
- Ten-minute after-dinner circadian-setting sniffari, usually meets up with neighborhood dogs to say hi and socialize.
- Dims home lights, closes drapes.
- Bedtime: turns off TV and router, brushes Charm's teeth, gives Charm a light, calming body massage (and head-to-toe scan), turns off all lights.

What does your forever tomorrow look like?

Epilogue

According to emerging research, the psychological benefits of owning a dog are real—and compelling. The skyrocketing adoption rate of "pandemic dogs" in the United States (adoptions were up 30 percent in 2020, according to the nonprofit database Shelter Animals Count) is a vivid testament to our collective confidence that dogs improve our mental health and mitigate loneliness. What we've known anecdotally for millennia is now borne out by the research: Study after study shows that having a pet helps you maintain an optimistic outlook on life and lessens the symptoms of depression and anxiety. Dogs are good for the soul—and for our health. The author and animal expert Karen Winegar says it perfectly: "The human-animal bond bypasses the intellect and goes straight to the heart and emotions and nurtures us in ways that nothing else can." Yes, indeed. Our dogs nourish our souls in ways that nothing else can. They enrich our lives so much, which is why losing a beloved pet can hurt as much as, or more than, losing a human family member or friend.

They've given so much to *us*. It's our turn to give to *them*. We hope this book inspires people to give their very best to their dogs. When we make the commitment to care for an animal, we assume a moral responsibility to do right by her. We take a silent oath to be a good dog parent. We *want* to do right by our dogs. We want our precious, furry new companion to be happy, fulfilled, and healthy—to fully express her doggie self. The best way to do this is to create an environment and to provide sustenance that stimulates and nourishes her body, her brain, and her soul.

We implore you not to kick yourself or feel guilt about your "old ways." The pet food industry has been extraordinarily successful in convincing dog guardians that their highly processed, nutritionally vapid fare is all your dog needs to live a healthy, long life. *Now you know better.*

Having read this book, you are fully empowered with the science (the why) and the tools (the how) to craft a Forever Dog wellness plan that works for you and that will set your dog on a transformational wellness journey. You do not need a wholesale overhaul; incremental changes deliver powerful results. You needn't break the bank; those organic blueberries in your fridge will give Fido a potent punch of antioxidants that speak to his genome. Our goal with this book is to provide you with the information you need to make the right choices for you and your pup; when you know better, you do better. Keep this book and visit online for updates and reference lists as you plan meals and activities for your pooch; and don't forget to mix it up. Variety is the spice of life—and so important in building a Forever Dog.

It may feel intimidating right now to know where to start, and you may doubt your own ability to get this right. Don't be dissuaded. We promise you: You can do this! It's really *not* that hard. It does take commitment and time, but we know you are committed, because you are about to finish this book! We also promise that the day will come when your confidence catches up to your knowledge. In the meantime, keep fighting the good fight. Keep reading (labels, articles, books), researching (there's so much available online), and engaging more deeply with the animal wellness community—there are legions of us around the globe; you need only look and you'll find the supportive community that resonates with you. In time, and with practice, your apprehension will diminish and your confidence will grow. Some things will work well and become part of your forever routine; other things you'll try and then move along. This is exactly how it's supposed to work.

The best part: One day, you'll simply *know*. You'll know you're doing right by your dog, because your dog will tell you. You'll see her energy, her luster, her vitality, her improved health, a new spring in her step, a sparkle in her eye—and you'll know. You'll know that *you* gave her the incredible gift of health and longevity. There's no feeling in the world more satisfying than knowing you've done your part to create a Forever Dog. Well done, guardian.

We wish you the best. Forever.

Acknowledgments

We have so many people to thank, so many people for whom we are wildly, obscenely, and humbly grateful. *The Forever Dog* has been a joyously collaborative, cooperative effort, with dozens of people generously sharing their insights, time, and expertise to make this book what it is meant to be: a revolution in global canine health. We have interviewed, connected with, partnered with, and learned from amazing people who have supported this project from its inception, starting with the world-class experts and scientists included in these pages. Throughout our project, their excitement about this book and their enthusiasm to share their research—indeed, their fundamental goal of helping dogs live longer, healthier lives—were a tremendous inspiration to us. It goes without saying that meeting some of the oldest dogs in the world, and their amazing owners, was an astonishing, once-in-a-lifetime gift we will never forget. Thank you, Cindy Meehl, for introducing me (Rodney) to Joni Evans, who suggested I talk to Kim Witherspoon, who ultimately guided us through this entire book-creating experience. We are appreciative of all things HarperCollins: Brian Perrin worked diligently with Kenneth Gillett's and Mark Fortier's teams, along with our own Rachel Miller, Marc Lewis, and Bea Adams, to coordinate the challenging task of releasing a book during a pandemic. Karen Rinaldi, at Harper Wave, masterfully orchestrated the entire project and introduced us to our biggest asset throughout this adventure, our science writer Kristin Loberg. Thank you, Kristin, for keeping track of hundreds of references and interviews and helping us organize the science Rodney amassed over several years. Assembling such a huge volume of information for such a wide audience was nothing short of overwhelming. You promised us it would come together, and it did.

We are so grateful we did this together. And to Bonnie Solow, who provided invaluable editorial suggestions to elevate the manuscript. Aunt Jo (Dr. Sharon Shaw Elrod), Steve Brown, Susan Thixton, Tammy Akerman, Dr. Laurie Coger, Sarah Mackeigan, Jan Cummings, and my (Karen's) BFF, Dr. Susan Recker, we appreciate your editorial suggestions. We are humbled to be part of this community of exceptionally talented experts committed to pet health. Renee Morin, your unwavering support of our Inside Scoop online community of 2.0 pet parents makes everyone feel welcome, and your behind-the-scenes help is priceless. Many other people were swift to offer help along the way, including Niki Tudge, Whitney Rupp, the entire Planet Paws team, and of course our always-faithful family members. All of the healthy, delicious, homemade meals provided by our mamas, Sally and Jeannine, during the creation of this book allowed us to keep working *and* be well nourished! Thank you, Mama Becker, for not only making Dr. Becker's Bites treats, but for making all of my animals' meals during this writing process, when I (Karen) was too busy. And to Annie (Karen's sister), who stayed up multiple nights editing, editing, editing: your suggestions were invaluable for clarity and cohesiveness. We also feel a deep sense of gratitude for our animal-loving wellness community and the thousands of Longevity Junkies around the world who asked for this book. Our international network of knowledgeable animal wellness advocates grows by the day; for all of you implementing Forever Dog principles with astounding results, your unwavering support and inspiring testimonies fuel our souls and our life's mission. Finally, we would be remiss not to conclude this book the same way it started, expressing our deepest gratitude to the dogs in our lives who have been our most powerful teachers and our closest friends. Our dogs make us better humans. Our hope is that this book makes us all better guardians.

Appendix

Testing Recommendations

Annual exams are important for health. Because dogs age much faster than humans, starting mid-life, I see many of my patients every six months to make sure we are updating their wellness protocols as aging (or a new symptom) occurs. Wellness is a dynamic process that requires ongoing modifications to a patient's diet and personalized health strategies to achieve our goal of maximizing health span. In addition to a complete physical examination, basic lab work (including complete blood count [CBC]and blood chemistry profile), fecal parasite test, and urinalysis are important components of your dog's annual exam. There are some additional diagnostics that can be useful in determining health status and how well your dog is aging. They can further help you stay ahead of any fomenting illness or disease in your pet:

➤ **Vitamin D test**—Dogs and cats require dietary vitamin D because they can not convert it from sunlight. However, both commercial and homemade pet food diets pose their own issues since the synthetic version in commercial dog food is difficult to absorb, and homemade diets are typically lacking it. Ask your veterinarian to add Vitamin D testing to routine blood work. Low vitamin D levels negatively impact your dog in many ways, including compromising their immune response.

➤ **Dysbiosis test**—This is one of the most important tests for your pet because upwards of 70 percent of the immune system is in the gut. Gut issues can lead to malabsorption, maldigestion, and in the end, a malfunctioning immune system. Restoring good gut health is imperative, especially for older and constantly ill pets.

➤ **C-reactive protein (CRP)**—When looking for system-wide inflammation, this test is the most attuned. It's now offered by most vets right in the office.

- **Cardiac biomarkers (brain natriuretic peptide; BNP):** A simple blood test can measure substances the heart releases when the organ is damaged or stressed. It's a great screening test for myocarditis, cardiomyopathies, and heart failure.
- **A1c:** Originally used as a tool to monitor diabetes, human biohackers, metabolomics researchers, and functional medicine practitioners started using A1c as a marker of metabolic health about a decade ago. A1c is actually an advanced glycation end product (AGE); it's a measurement of how much hemoglobin (a protein that carries your oxygen) is covered in sugar (glycated). The higher your A1c, the more inflammation, glycation, and metabolic stress you have. Same for your dog.
- **A combined tick-borne and heartworm illness test:** Gone are the days of a simple heartworm test in many places around the world, including North America. Ticks are everywhere and harbor potentially fatal diseases that are much more common than heartworm. Lyme and other tick-borne diseases are quietly becoming an epidemic in dogs and people in certain areas. You can ask your vet to screen for heartworm, Lyme disease, and two strains each of *Ehrlichia* and *Anaplasma* using the SNAP 4Dx Plus (from Indexx Labs) or the AccuPlex4 test (Antech Diagnostics).
- If your dog tests positive on one of these screening tests for Lyme disease, it means he's been exposed. **It doesn't mean he has Lyme disease.** In fact, research shows most dogs' immune systems do exactly what they're supposed to do and mount an immune response to the bacteria and eliminate it. But in about 10 percent of cases, dogs become infected and can't clear the spirochete. These dogs need to be identified and treated in a timely manner, before symptoms start. The test that differentiates Lyme exposure from Lyme infection/disease is called a Quantitative C6 (QC6) blood test. Do not let your vet prescribe antibiotics until the QC6 demonstrates your dog is currently positive for Lyme infection. If you use antibiotics for any reason, make sure you focus on the microbiome-building protocols in this book. One of these simple blood tests to screen for tick-borne diseases is recommended every six to twelve months (depending on how rampant these diseases are in your area and the potency and frequency of flea and tick pesticides applied). If you are using all-natural preventives, test more frequently; they're not as effective as the hard-core pesticides (but also not as toxic). If you're

using prescription flea and tick medications from your vet, do an AccuPlex4 or SNAP 4DX Plus every year, and detox!

Note: wwww.foreverdog.com has more info on innovative biomarkers, wellness diagnostics, tests, and labs.

Nutritional Analysis for Beef Meal

Grams	Pounds	Ounces	Percent	Ingredient
2,270.0	5.00	80.00	58.07%	beef, ground, 93% lean, 7% fat, crumbles, cooked, pan-browned
908.0	2.00	32.00	23.23%	beef liver, cooked, braised
454.0	1.00	16.00	11.61%	asparagus, raw
113.5	0.25	4.00	2.90%	spinach, raw
56.8	0.13	2.00	1.45%	sunflower seed kernels, dried,
56.8	0.13	2.00	1.45%	hempseed
25.0	0.06	0.88	0.64%	calcium carbonate
15.0	0.03	0.53	0.38%	cod liver oil, Carlson, 400IU / tsp
5.0	0.01	0.18	0.13%	ginger, ground
5.0	0.01	0.18	0.13%	kelp meal, seaweed Tidal Organics
3,909	8.61	137.76	100.00%	

MACRONUTRIENT ANALYSIS			
	Atwater Standard		
Composition	as formulated	DM	% kcal
Protein	25%	66%	54%
Fat	9%	23%	42%
Ash	2%	6%	
Moisture	63%		
Fiber	1%	2%	
Net Carbs	2%	4%	3%
Sugars (limited data)	0%	1%	1%
Starch (limited data)	0%	0%	0%
Total			100%

Macronutrient Information	
total kcal in recipe	7,098
kcal / oz	52
kcal per pound	824
kcal / day	342
recipe makes, # of days	20.7
kcal / kg	1,817
kcal per kg DM	4,863
grams to feed per day	188
ounces to feed per day	6.6

Desired Weight		10.0	Lbs							40.0				
		4.5	Kg							18.2				
Activity Level, FEDIAF 2016	k Factor	kcal / day	oz/day	g/day	% of wt	cpp	cpkg	unit/d	kcal / day	oz/day	g/day	% of wt	cpp
Adult													
Resting energy	70	218	4.2	120	2.6%	21.8	47.9	3.8	616	12.0	339	1.9%	15.4
Adult – Indoor sedentary	85	265	5.1	146	3.2%	26.5	58.2	4.7	748	14.5	412	2.3%	18.7
Adult – Less Active	95	296	5.7	163	3.6%	29.6	65.1	5.2	836	16.2	460	2.5%	20.9
Adult – Active	110	342	6.6	188	4.2%	34.2	75.3	6.0	969	18.8	533	2.9%	24.2
Adult – More Active	125	389	7.6	214	4.7%	38.9	85.6	6.9	1,101	21.4	606	3.3%	27.5
Adult – Very Active	150	467	9.1	257	5.7%	46.7	102.7	8.2	1,321	25.6	727	4.0%	33.0
Adult - Working Dog	175	545	10.6	300	6.6%	54.5	119.9	9.6	1,541	29.9	848	4.7%	38.5
Adult - Sled Dog	860	2,677	52.0	1,473	32.5%	267.7	589.0	47.2	7,572	147.0	4,167	23.0%	189.3

Minerals	AAFCO 2017 - Adult – Active				Daily Amt
	Unit	minimums	maximums	Recipe	
Ca	g	1.25	6.25/4.5	1.67	0.54
P	g	1.00		1.66	0.57
Ca: P ratio	:1	1:1	2:1	1 : 1	
K	g	1.50		2.27	0.78
Na	g	0.20		0.41	0.14
Mg	g	0.15		0.22	0.08
Cl (no USDA data)	g	0.30		0.01	0.00
Fe	mg	10.00		21.81	7.47
Cu	mg	1.83		19.02	6.51
Mn	mg	1.25		1.59	0.54
Zn	mg	20.00		30.74	10.53
I (no USDA data)	mg	0.25	2.75	0.475	0.16
Se	mg	0.08	0.50	0.124	0.04

Vitamins	AAFCO 2017 - Adult – Active				Daily Amt
	Unit	minimums	maximums	Recipe	
Vit A	IU	1,250.00	62,500	42940.13	14,704
Vit D	IU	125.00	750	252.63	87
Vit E	IU	12.50		12.90	4
Thiamine, B1	mg	0.56		0.73	0.3
Riboflavin, B2	mg	1.30		5.24	1.8
Niacin, B3	mg	3.40		46.56	15.9
Pantothenic Acid,B5	mg	3.00		11.95	4.1
B6 (Pyridoxine)	mg	0.38		2.91	1
Vit B12	mg	0.01		0.099	0.034
Folate	mg	0.05		0.432	0.148
Choline	mg	340.00		860.95	295

FATS	AAFCO 2017 - Adult – Active		per 1000 kcal		Daily Amt
	Unit	minimums	maximums	Recipe	
Total	g	13.80	82.5	47.06	16.11
Saturated	g			15.89	5.44
Monounsaturated	g			15.19	5.20
Polyunsaturated	g			7.11	2.43
LA	g	2.80	16.30	5.12	1.75
ALA	g			0.65	0.22
AA	g			0.44	0.15
EPA + DHA	g			0.41	0.14
EPA	g			0.18	0.06
DPA	g			0.09	0.03
DHA	g			0.23	0.08
omega-6/omega-3	:1		30:1	5.25	

Amino Acids	AAFCO 2017 - Adult – Active		per 1000 kcal		Daily Amt
	Unit	minimums	maximums	Recipe	
Total protein	g	45.00		135.74	46.48
Tryptophan	g	0.40		0.99	0.34
Threonine	g	1.20		5.26	1.80
Isoleucine	g	0.95		5.98	2.05
Leucine	g	1.70		10.84	3.71
Lysine	g	1.58		10.69	3.66
Methionine	g	0.83		3.40	1.17
Methionine - cystine	g	1.63		5.08	1.74
Phenylalanine	g	1.13		5.69	1.95
Phenylalanine - tyrosi	g	1.85		10.06	3.44
Valine	g	1.23		6.97	2.39
Arginine	g	1.28		9.09	3.11

Red-shaded areas (if any) do not meet dog growth > of EU, AAFCO.

Nutritional Analysis for Turkey Meal
with Supplements

Turkey Dog Food Recipe

RECIPE INGREDIENTS

Item	Grams	Pounds	Ounces	Percent
turkey, ground, 85% lean, 15% ft, pan-broiled, crumbles	2,270.00	5.00	80.07	51.23%
beef liver, cooked, braised	908.00	2.00	32.03	20.49%
Brussels Sprouts, Cooked, Boiled, Drained W/O Salt	454.00	1.00	16.01	10.25%
Beans, Snap, Green, Frozen, All Styles, Unprepared	454.00	1.00	16.01	10.25%
Endive, Raw,	227.00	0.50	8.01	5.12%
Salmon Oil, Wild Slmn Oil Blend, Omega Alpha	50.00	0.11	1.76	1.13%
Calcium Carbonate	25.00	0.06	0.88	0.56%
Vitamin D3, 400IU/G	3.00	0.01	0.11	0.07%
Potassium Solaray, 99 Mg/Cap, 1 G= 1 Cap	25.00	0.06	0.88	0.56%
Magnesium Citrate, 200 Mg / Tablet 1 G = 1 Tablet	3.00	0.01	0.11	0.07%
Manganese Chelate --10 Mg	1.00	0.00	0.04	0.02%
Zinc -- Nature'S Made, 30Mg Tablet	4.00	0.01	0.14	0.09%
Iodine, Whole Foods, 360 Mcg/Cap	7.00	0.02	0.25	0.16%
Vitamin E 400 IU, 1gm = 1 cap, Bluebonnet	0.13	0.00	0.00	0.00%
Total	**4,431.13**	**9.77**	**156.30**	**100.00%**

MACRONUTRIENT ANALYSIS

Nutrient content of natural foods vary, sometimes significantly.
Use the nutrient content numbers as approximations only.

Composition	As Formulated	DM	% kcal
Protein	19.33%	54.01%	39.78%
Fat	11.23%	31.36%	56.1%
Ash	2.52%	7.05%	
Moisture	64.2%		
Fiber	0.71%	1.99%	
Net Carbs	2%	5.59%	4.12%
Sugars (limited data)	0.24%	0.67%	0.49%
Starch (limited data)	0.16%	0.44%	0.32%
Total			**100%**

MACRONUTRIENT INFORMATION

total kcal in recipe	7,538.38
kcal / oz	48.23
kcal per pound	771.67
kcal / day	2,068.33
recipe makes, # of days	3.64
kcal / kg	1,701.20
kcal per kg DM	2,108.97
Amount to Feed per Day (gm)	1,215.80
Amount to Feed per Day (oz)	42.89
keto ratio (g fat/ (g protein + g net carb))	0.53

MINERALS

	Unit	Min	Max	Recipe	Daily Amt
Ca	g	1.25	0.00	1.54	3.19
P	g	1.00	4.00	1.45	3.00
Ca: P	ratio	1 : 1	2 : 1	1.06 : 1	
K	g	1.25	0.00	1.79	3.70
Na	g	0.25	0.00	0.37	0.77
Mg	g	0.18	0.00	0.22	0.45
Cl (no USDA data)	g	0.38	0.00	0.00	0.00
Fe	mg	9.00	0.00	15.33	31.71
Cu	mg	1.80	0.00	17.85	36.92
Mn	mg	1.44	0.00	2.16	4.47
Zn	mg	18.00	71.00	33.62	69.53
I (no USDA data)	mg	0.26	0.00	0.33	0.69
Se	mg	0.08	0.14	0.15	0.32

VITAMINS

	Unit	Min	Max	Recipe	Daily Amt
Vit A	IU	1,515.00	100,000.00	39,965.19	82,661.27
Vit C	mg	0.00	0.00	12.02	24.85
Vit D	IU	138.00	568.00	242.30	501.16
Vit E	IU	9.00	0.00	9.00	18.62
Thiamine, B1	mg	0.54	0.00	0.62	1.29
Riboflavin, B2	mg	1.50	0.00	5.03	10.41
Niacin, B3	mg	4.09	0.00	45.14	93.37
Pantothenic Acid, B5	mg	3.55	0.00	13.10	27.10
B6 (Pyridoxine)	mg	0.36	0.00	2.75	5.70
Vit B12	mg	0.01	0.00	0.09	0.19
Folic Acid	mg	0.07	0.00	0.41	0.86
Choline	mg	409.00	0.00	749.15	1,549.50
Vit K1 (minimal data)	mg	0.00	0.00	158.03	326.87
Biotin (minmal data)	mg	0.00	0.00	0.00	0.00

AMINO ACIDS

	Unit	Min	Max	Recipe	Daily Amt
Total protein	g	45.00	0.00	113.65	235.07
Tryptophan	g	0.43	0.00	1.32	2.72
Threonine	g	1.30	0.00	5.00	10.34
Isoleucine	g	1.15	0.00	5.08	10.51
Leucine	g	2.05	0.00	9.56	19.78
Lysine	g	1.05	0.00	9.55	19.76
Methionine	g	1.00	0.00	3.16	6.54
Methionine - cystine	g	1.91	0.00	4.61	9.53
Phenylalanine	g	1.35	0.00	4.83	9.99
Phenylalanine - tyrosine	g	2.23	0.00	8.91	18.43
Valine	g	1.48	0.00	5.70	11.80
Arginine	g	1.30	0.00	7.65	15.83
Histidine	g	0.58	0.00	3.33	6.89
Purines	mg	0.00	0.00	0.00	0.00
Taurine	g	0.00	0.00	0.02	0.05

FATS

	Unit	Min	Max	Recipe	Daily Amt
Total	g	13.75	0.00	66.00	136.52
Saturated	g	0.00	0.00	15.85	32.79
Monounsaturated	g	0.00	0.00	19.03	39.35
Polyunsaturated	g	0.00	0.00	15.42	31.90
LA	g	3.27	0.00	12.96	26.80
ALA	g	0.00	0.00	0.76	1.56
AA	g	0.00	0.00	0.69	1.42
EPA + DHA	g	0.00	0.00	2.12	4.38
EPA	g	0.00	0.00	1.28	2.64
DPA	g	0.00	0.00	0.04	0.08
DHA	g	0.00	0.00	0.84	1.74
omega-6/omega-3	ratio			4.75 : 1	

Twenty Questions to Ask a Prospective Breeder

GENETIC AND HEALTH SCREENING TESTS

1. Have all of the currently appropriate DNA tests for the breed been conducted on the dam (mom)? (Find a list by breed at www.dogwellnet.com.)

2. Have all of the currently appropriate DNA tests for the breed been conducted on the sire (dad)?

3. What were the results of the Orthopedic Foundation for Animals (OFA) hip dysplasia (or PennHip), elbow, and patella screening results for both parents?

4. For affected breeds, when were the dam's and sire's thyroid results last registered with the OFA thyroid database?

5. If indicated for breed, have the dam's and sire's eyes been evaluated by an ophthalmologist and results reported to Companion Animal Eye Registry (CERF) or OFA?

6. Are there any breed-related issues the breeder is trying to address/rectify/improve through mating this pair?

EPIGENETICS

7. What percentage of the dam's and sire's diet is unprocessed or minimally processed food?

8. What are the dam's and sire's vaccine protocols?

9. Is the puppy's vaccination protocol determined by a nomograph (testing of the mom's antibody levels to determine what day vaccines will be effective for the puppies)?

10. How often are pesticides applied to the parents (topical or oral heartworm, flea, and tick medications)?

SOCIALIZATION, EARLY DEVELOPMENT, AND WELLNESS

11. What early socialization program (day 0–63) does the breeder institute prior to placing puppies in their new homes?

12. Does the breeding contract require that puppies be spayed or neutered by a certain age?

13. If yes, does the sterilization clause include options for vasectomy or hysterectomy?

14. Does the contract require you to attend training classes with your pup?

15. If appropriate for the breed, are puppies' eyes checked by a veterinary ophthalmologist between six and eight weeks of age?

16. Have the puppies had a basic checkup by the breeder's regular veterinarian prior to going to new homes, and at what age will puppies be released?

TRANSPARENCY

17. Will the breeder allow you to visit their home or facility (in person or via live video) and provide references for you to call?

18. In the event you cannot keep your puppy or things don't work out, will the breeder take the puppy back at any time?

19. Is the breeder (or someone in their network) available for support should you need it?

20. Will your puppy packet include all of the following?

     » Contract

     » AKC or applicable registration application or certificate if already registered

     » Other breed registry registration, if applicable (i.e., Australian Shepherd Club of America)

     » Litter pedigree

     » Copies of puppy eye exam findings, if applicable

     » Puppy general health summary from veterinarian (medical record from initial veterinary visit)

     » Dam health clearances, including copies of DNA results

     » Sire health clearances, including copies of DNA results

     » Photos of sire and dam

     » Educational resources (suggested feeding schedule, suggested vaccine protocols and suggested dates for antibody titers to ensure immunization occurred, training resources)

RAW BONE RULES

Eating crunchy granola doesn't remove plaque from your teeth, and feeding dogs crunchy treats doesn't remove plaque from their teeth, either. Yet people still believe dog biscuits "clean" their dog's teeth. They don't! This is a shameless marketing ploy. There are three ways to remove plaque from your dog's teeth: your vet can professionally clean them (this is the most effective way to get a clean mouth, but it usually involves anesthesia); you can brush them every night after dinner (which we're fans of); and you can encourage your dog to participate in the removal of plaque via the act of chewing (aka "mechanical abrasion"). When your dog chews on a raw recreational bone, especially a meaty one with cartilage and soft tissue still attached, his teeth get the equivalent of a good brushing and flossing, but he's doing the work, not you. One study found offering dogs raw bones removed the majority of plaque and tartar from the molars and the first and second premolars in less than *three days*! They're called recreational bones because dogs love to gnaw on them, but they aren't meant to be chewed up and swallowed. And they come with a long list of rules.

You should be able to find a selection of raw bones in the freezer section of your neighborhood independent pet store, with knowledgeable staff to help you choose the correct size bone for your dog. Outside of independent pet stores, supermarket meat counters and butcher shops should carry raw (not steamed, smoked, boiled, or baked) knucklebones. If you're having trouble finding them, they are sometimes referred to as soup bones and can be found in the refrigerated or frozen food section. To maintain freshness, keep the bones in the freezer and thaw one at a time to give to your pup. Generally speaking, knucklebones from large mammals (beef, bison, venison) are the safest options. Other tips:

➤ Match the bone size to your dog's head. There's really no such thing as a too-big bone, but there are definitely bones that are too small for some dogs. Too-small bones can be choking hazards and can also cause significant oral trauma (including broken teeth).

➤ If your dog has had restorative dental work or crowns, or if your dog has fractured teeth or soft teeth (very old dogs), do not offer recreational bones.

➤ Remember to keep a watchful eye on your dog at all times when he has a bone. Under no circumstances should your pet be alone with a bone.

➤ In multidog households, separate your dogs before giving recreational bones, to keep the peace. This goes for dogs that casually know each other or are close friends. Resource guarders should not be offered raw bones. Collect bones at the end of the chewing session (fifteen minutes is a good time frame, for starters).

➤ A word to the wise for those incorporating bone marrow into their dog's diet: it is a fatty substance and will add to your dog's daily caloric intake. With this in mind, pets with pancreatitis should not eat bone marrow. If your pet has a sensitive stomach there are ways to incorporate marrow into their diet, like scooping it out until your dog has adjusted to the additional fat in their diet or pacing chewing sessions for shorter amounts of time, say fifteen minutes per day. Without these precautions in place, you run the risk of stomach upset and even diarrhea.

➤ One last word to the wise: raw bones get messier the more your dog chews on them. Try offering the bone outside or somewhere that is easy to clean with hot, soapy water.

➤ To make sure your dog is getting the right size bone, use their head for reference. No bone will be too big, but if it is too small it may be a choking hazard or cause oral trauma such as a broken tooth.

Additional Resources

Check www.foreverdog.com for current updates.

RESOURCES TO FIND A REHABILITATION PROFESSIONAL

➤ Graduates of the Canine Rehabilitation Institute: www
.caninerehabinstitute.com/Find_A_Therapist.html
➤ Canadian Physiotherapy Association: www.physiotherapy.ca
/divisions/animal-rehabilitation
➤ Online directory of the American Association of Rehabilitation
Veterinarians: www.rehabvets.org/directory.lasso
➤ Graduates of the Canine Rehabilitation Certificate Program: www
.utvetce.com/canine-rehab-ccrp/ccrp-practitioners

TRAINING AND BEHAVIOR RESOURCES

➤ Certification Council for Professional Dog Trainers (CCPDT): www
.ccpdt.org
➤ International Association of Animal Behavior Consultants (IAABC):
www.iaabc.org
➤ Karen Pryor Academy: www.karenpryoracademy.com
➤ Academy for Dog Trainers: www.academyfordogtrainers.com
➤ Pet Professional Guild: www.petprofessionalguild.com
➤ Fear Free Pets: www.fearfreepets.com
➤ American College of Veterinary Behaviorists: www.dacvb.org

EARLY PUPPY PROGRAMS

➤ Avidog: www.avidog.com
➤ Puppy Culture: www.shoppuppyculture.com
➤ Enriched Puppy Protocol: https://suzanneclothier.com/events
/enriched-puppy-protocol/
➤ Puppy Prodigies: www.puppyprodigies.org

CONCIERGE WELLNESS SERVICES THAT EMBRACE
FUNCTIONAL VETERINARY MEDICINE

➤ College of Integrative Veterinary Therapies: www.civtedu.org
➤ American Veterinary Chiropractic Association: www.animal chiropractic.org
➤ International Veterinary Chiropractic Association: www.ivca.de
➤ American College of Veterinary Botanical Medicine: www.acvbm .org
➤ Veterinary Botanical Medicine Association: www.vbma.org
➤ Veterinary Medical Aromatherapy Association: www.vmaa.vet
➤ American Academy of Veterinary Acupuncture: www.aava.org
➤ International Veterinary Acupuncture Society: www.ivas.org
➤ International Association of Animal Massage and Bodywork: www .iaamb.org
➤ American Holistic Veterinary Medical Association: www.ahvma.org
➤ Raw Feeding Veterinary Society: www.rfvs.info

"Supplemental Feeding" Dog Foods

All pet foods in the United States must make a nutritional claim on their packaging. If you live in Canada or other countries where there aren't label claim regulations, unfortunately, it's up to you to do your research to assess if the food you are buying is nutritionally adequate. In the United States, a label that reads "For supplemental or intermittent feeding" means the food is nutritionally incomplete—it's deficient in critical vitamins and minerals that must be supplied in the dog's diet but aren't supplied in that food. Regardless of where you live, if commercial pet food labels don't come with a statement of nutritional adequacy and the company can't provide a complete nutritional analysis (as compared to AAFCO, NRC, or FEDIAF), you should assume the food does not meet your dog's daily nutritional requirements. As long as you've considered processing temperature, purity, and sourcing of all the ingredients, these foods can be great dollops, treats, toppers, or short-term temporary diets for adult dogs (one out of seven or two out of fourteen meals). These incomplete diets are not meant to be fed consistently as a sole food source;

the problem is, they are. And they cause all sorts of roadblocks on a dog's path to longevity.

When dogs are deficient in critical vitamins and minerals that function as cofactors for key enzymatic reactions and facilitate the production of key proteins, the body does not function optimally on a cellular level, which leads to metabolic and physiologic stress, over time. Eventually, disease is inevitable. The problem is you can't see outward signs of these micronutrient deficiencies until your dog's body is so depleted there's no chance at the Guinness World Records in her future. People are so confused about what to feed, and these days often with serious economic constraints. We totally get it. This scenario has provided a ripe opportunity for pet food companies to offer a timely solution: much cheaper, fresher, unbalanced dog food (compared to many other raw or minimally processed brands that have intentionally created well-formulated, nutritionally complete, and therefore more expensive diets). For 3.0 super-knowledgeable pet parents who aren't afraid of math, some of the commercially available raw food "grinds" (a blend of meat, bone, and organ) or other cooked, dehydrated, or freeze-dried "base mixes" of meats and veggies provide a good option; these mixes can be a godsend on the wallet. These are fine to include as a topper or treat (less than 10 percent of your dog's calories for the day).

If you want to use these unbalanced dog food blends as a base for your dog's daily meals, you'll have to fill in all the blanks, nutritionally speaking. There are many micro-companies making small-batch dog foods that have the potential to be great food for your dog, if they're balanced by you (yes, with a calculator). Transparent companies have downloadable PDFs on their websites showing the lab nutritional analyses of their (incomplete) diets. This info can be plugged into a raw food spreadsheet to compare to current accepted nutritional standards (found at www.foreverdog.com) to determine what nutrients need to be added. Many 3.0 pet parents in our community do this. It's a fantastic way to feed balanced, fresh food on a shoestring budget. The cheapest way to feed a 100 percent human-grade, nutritionally optimal diet to your dog is to shop power sales, join a co-op, buy in bulk, and make food yourself following a nutritionally complete recipe, at home. This just isn't feasible for so many people we know.

If you attempt to balance unbalanced commercial dog food diets on your own, your dog is relying on you to identify not only what nutrients are missing but what amounts need to be added to at least meet his minimum daily requirements. "Supplemental feeding" diets aren't bad; in fact, with astute evaluation and appropriate execution in the bowl, they can be *quite beneficial* for your wallet, depending on the intensity of processing and the quality of raw ingredients. But for this category of pet food, one thing matters more than anything: the ethics of the company. How transparent are they about sharing nutritional analysis test results for their products? This is a critical question we recommend asking prior to feeding these foods for any length of time (more than a CLT or an occasional light meal). There are "supplemental feeding" dog food brands (usually those that don't make any nutritional claim statement fall into this category) across most pet food categories, including lots of raw food brands and several gently cooked foods from micro-companies you can buy at farmers' markets, pet boutiques, big box stores, and online. You can probably already see what some minuses are with this category . . .

Most companies producing "prey model" grinds—"80/10/10" (meat/bone/organ) base mixes, raw food components, or "ancestral dog food"—don't supply a list of what ingredients or supplement amounts you'll need to add to nutritionally balance their unbalanced diets. Even worse, it can be hard to impossible even to get the raw data you need from their customer service department to attempt the calculations to fix the deficiencies. **Some companies selling you dog food won't provide information about what's in their food**. Scary, because either they don't know the nutritional profile of their food or they don't want you to know. In fact, some companies tout that if dog owners just rotate through all of their flavors or proteins, they'll meet minimum nutritional requirements over time, without ever providing proof of this being a viable way of meeting your dog's nutrient requirements. This makes veterinarians very angry because it's usually not true. **The problem with unknowingly rotating through a large variety of unbalanced diets is that your dog remains nutritionally deficient.** This is one of the biggest reasons we see animals eating less-processed, fresher, or raw foods do poorly: their diet is fresh. But deficient.

Be leery of feeding foods from companies offering vague nutrition recommendations to balance their product, such as "add kelp and omega-3s to balance our diets." "Rotation over time" (feeding a rotation of several different meats, bones, and organs) is another nutrition concept that drives vets crazy because very few people or companies can demonstrate they're actually meeting trace nutrient requirements in any capacity. This is a bigger problem than most people realize, and many vets are already frustrated with their clients' experimenting with "alternative, nontraditional food categories" and veering away from ultra-processed, highly refined "food-like particles" (what fresh-feeding veterinarian Dr. Ian Billinghurst calls kibble). If you don't know in your heart that you can assure your dog is getting everything she needs, nutritionally, at least most of the time, feed these unformulated commercial diets a couple of times a week (two out of fourteen meals) or daily, as a Core Topper (10 percent of calories). Professionals in the directory at www.freshfoodconsultants.org can help you balance these products, or you can use the www.petdietdesigner.com spreadsheet.

Many 3.0 pet parents also master feeding homemade variations of raw meaty bone diets (RMBDs) or bones and raw food diets ("BARF," as you'll see online) that meet minimum nutritional requirements. This style of feeding involves blending together a variety of meats, bones, glands, and organs to mimic prey, and can be done successfully to avoid imbalances by using one of the many raw food–balancing spreadsheets available.

Notes

Our selected list of notes to accompany statements made in the book became a tome in itself due to the volume of sources and scientific literature we could have cited. We moved them online to www.foreverdog.com, where we can also keep them updated. For general statements made in the book, we trust you can find a wellspring of references and evidence yourself online with just a few taps of the keyboard, assuming you visit reputable sites that post fact-checked, credible information that's been vetted by experts. This is especially important when it comes to matters of health and medicine. The best medical journal search engines that do not require a subscription, many of which are listed in the notes include: pubmed.gov (an online archive of medical journal articles maintained by the United States National Institutes of Health's National Library of Medicine); sciencedirect.com and its sibling SpringerLink link.springer .com; the Cochrane Library at cochranelibrary.com; and Google Scholar at scholar.google.com, which is a great secondary search engine to use after your initial search. The databases accessed by these search engines include Embase (owned by Elsevier), Medline, and MedlinePlus and cover millions of peer-reviewed studies from around the world. We've done our best to include all the studies specifically highlighted and added more in places to round out conversations. Use the entries as launchpads for further inquiry, and don't forget to check out our website at www.foreverdog.com for updates.

Index

Page references in *italics* refer to illustrations.

About the Authors

Rodney Habib is a multiple award-winning blogger, public speaker, and the founder of Planet Paws, the most popular pet health page on Facebook, and the creator of the documentary *The Dog Cancer Series: Rethinking the Canine Epidemic.* His educational online video posts have been viewed and shared millions of times the world over, earning him a reputation as one of the most sought-after, engaging authorities on pet health.

Veterinarian **Karen Shaw Becker**'s deliberate, commonsense approach to creating vibrant health for companion animals has been embraced by millions of pet lovers around the world, making her the most followed vet on social media. She has spent her career as a small-animal clinician and writes and lectures extensively. Dr. Becker also serves as a wellness consultant for a variety of health-oriented organizations.

READ MORE BY
DR. KAREN SHAW BECKER
& RODNEY HABIB